American Blacksmithing
and
Twentieth Century Toolsmith and Steelworker

American Blacksmithing
and
Twentieth Century Toolsmith and Steelworker

- *Both complete in one volume*
- *Techniques and mechanics of the craft with descriptive line drawings*

by Holmstrom and Holford

Greenwich House
New York

Copyright © 1982 by Crown Publishers, Inc.

All rights reserved.

This 1982 edition is published by Greenwich House, a division of Arlington House Inc., distributed by Crown Publishers, Inc.

Manufactured in the United States of America

Library of Congress Cataloging in Publication Data
Holmstrom, John Gustaf
   American blacksmithing, toolsmiths' and steelworkers' manual.

   Includes index.
   1. Blacksmithing. 2. Steelwork. I. Holford, Henry, 1876- .
II. Title.
TT220.H67   1982   682   82-9240
AACR2
ISBN : 0-517-390485

h g f e d c b a

FOREWORD

Americans have experienced a resurgence of interest in old crafts and techniques, intrigued by the quality and individualism of handmade objects. Mass production increased efficiency and speed in manufacturing, but its standardized products lacked the character and diversity of items made by master craftsmen.

American Blacksmithing is a clear, practical guide to the art of blacksmithing, a craft that is more complex than just shoeing horses. It will appeal to anyone who appreciates fine craftsmanship, either in the abstract or to anyone who wants to practice the skill in actuality. *American Blacksmithing* covers the tools of the trade and their uses—the anvil, hammer, tool table, the blacksmith's fire—as well as giving instructions on how to shoe horses, mules, oxen, how to make plowshares, babbitting, and other aspects of the blacksmithing trade.

The book also discusses toolsmithing and steelworking, explaining how to make the tools necessary for blacksmithing and other tools used in ironworking. The descriptions are definitely approached in a "hands on" manner—these techniques are not discussed just in theory, they are geared for the reader's use.

American Blacksmithing provides all the information required by either the blacksmithing enthusiast or the armchair craftsman who would like to learn more about this demanding, but rewarding, craft. Working with iron is not just part of our past, it is an exciting element of the current craft movement in America.

American Blacksmithing

PREFACE

WHAT prompted the author to prepare this book was the oft-repeated question, by blacksmiths and mechanics of all kinds, as well as farmers: "Is there a book treating on this or that?" etc., etc. To all these queries I was compelled to answer in the negative, for it is a fact that from the time of Cain, the first mechanic, there has never been a book written by a practical blacksmith on subjects belonging to his trade. If, therefore, there has ever been such a thing as "filling a long-felt want," this must certainly be a case of that kind.

In medicine we find a wide difference of opinion, even amongst practitioners of the same school, in treating diseases. Now, if this is so where there is a system, and authority for the profession, how much more so must there be a difference of opinion in a trade where every practitioner is his own authority. I shall, therefore, ask the older members of the blacksmith fraternity to be lenient in their judgment if my ideas don't coincide with theirs. To the apprentice and journeyman I would say: do as I do until you find a better way.

The author has been eminently successful in his practice, and his ideas have been sought by others wherever he has been, blacksmiths coming even from other States to learn his ways.

This little book is fresh from the anvil, the author taking notes during the day while at work, compiling the same into articles at night.

He is indebted to a number of writers for articles in this book treating on subjects belonging to their trades, in which they have been regarded as experts.

CHAPTER I

THE SMITH

FOR centuries the blacksmith has been a prominent person, and it is natural he should have been, when we consider the variety of work he had to do. From the heavy axle and tire, down to the smallest rivet in the wagon, they were all made by the smith. Bells and bits as well as the ornamental parts of the harness, they were all made by the smith. From the crowbar and spade down to the butcher and pocket knife, they were all made by the smith. The carpenter's tools, from the broadax and adz down to the divider and carving steel, they were all made by the smith. From the heavy irons in the fireplace down to the frying-pan and locks on the kitchen doors; knives and forks on the dining-table, they were all made by the smith. From the gun on the shoulder of the soldier and the saber in the hands of the officer, the spurs and pistol for the commander, they were all made by the smith. From the heavy anchor and its

chain to the smallest pulley in the rigging of the ship, they were all made by the smith.

From the weather vane on the church spire, and

THE SMITH

the clock in the tower down to the lock of the doors and the artistic iron cross over the graves in the church yard, they were all made by the smith. No wonder, then, that the smith was respected. Vulgar people swear by the devil, religious by the saints, but the Swedes (the makers of the best iron) prefer to swear by the smith. The smith was a well-liked person in society, respected and even admired for his skill, his gentlemanly behavior and good language. His stories and wit were the sole entertainment in many a social gathering. Things have changed in the last few decades. Most of the articles formerly made by the smith are now manufactured by machinery, and the respect for the smith is diminished in the same proportion. Not because there is not enough of the trade left to command respect—there is yet more left than any man can successfully learn in a short lifetime. But it has made it possible for men with less training and ability to enter the trade and consequently lower the standing of the smith. The result is, that there is a complaint that the smith is not esteemed as formerly, and I have been inclined to join in the lamentation. But instead of doing this I shall ask my brother smiths to unite with me in an effort to elevate the craft.

THERE ARE SMITHS AND SMITHS

I have had the pleasure of becoming acquainted with a great number of intelligent and respected smiths. People that did not know them would ask: "What is he?" and when informed that he is a blacksmith would say: "He doesn't look it; I thought he was a business man"; another, "He looks like a lawyer or a minister." From this you will understand how, in many cases, the blacksmith looks. A great preacher was announced to preach in a neighboring town, and I went to hear him. Just as I sat down in the pew one of the local smiths walked up to me and sat down by my side. He was a *blacksmith* and he "looked it." Under his eyes was a half moon in black; on both sides of his nose was a black stripe that had been there since his first day in the shop. His ears, well, you have seen a clogged-up tuyer iron. His clothes were shabby and his breath a strong mixture of tobacco and whisky, which made wrinkles on the nose of the lady in front of us. I was somewhat embarrassed, but the sermon began. As the congregation arose, I opened the hymnbook and my brother smith joined, and with a hand that looked like the paw of a black bear, he took hold of the book.

After service I was invited by the smith to dinner. Between a number of empty beer kegs we managed to reach the door of the house and everything inside looked the color of his trade. I looked around for books and other articles of culture and found a hand organ and a pack of cards. The only book or reading matter to be found was a weekly of the kind that tells of prize fights, train robberies and murder. I had a fair dinner and told my host that I had to start for home. By this time I was sick of his language—profanity, mixed with a few other words—and I started to leave. On my way to the livery stable I passed my friend's shop, and he said it would not be fair to leave before I had seen his shop. "I have," said he, "a very good shop." The shop was a building of rough boards 18x20—the average farmer has a better wood shed. A big wood block like the chopping block in a butcher shop, was placed so close to the forge that he could only get edgewise between. On this block was to be found, anvil and all his tools, the latter were few and primitive, and would have been an honor to our father Cain, the first mechanic and blacksmith. What thinkest thou, my brother smith? Having spent years to learn the trade you must submit to a comparison with smiths of this caliber. Their work being inferior they must work cheap, and in some, perhaps many, cases set the price on your work. Smiths of this kind cannot expect to be respected. There might be some show for them in Dawson City or among the natives in that vicinity, but not in civilized America.

INTEMPERANCE

ONE of the chief reasons why the blacksmith is not so successful nor respected as before is his intemperance. The danger for the smith becoming a drunkard is greater than for any other mechanic. It is often the case that when a customer pays a bill the smith is requested to treat. This is a bad habit and quite a tax on the smith. Just think of it—fifteen cents a day spent for liquor, will, in twenty-five years, amount to $9,000. Then add to this fifteen cents a day for cigars, which will, in twenty-five years, amount to $9,000 at ten per cent compound interest. If these two items would be saved, it will give a man a farm worth $18,000 in twenty-five years. How many smiths are there who ever think of this? I would advise every one to put aside just as much as he spends for liquor and tobacco; that is, when you buy cigars or tobacco for twenty-five cents put aside as much. When you buy liquor for one dollar put aside one dollar. Try this for one year and it will stimulate to continual effort in that direction. The best thing to do is to "swear off" at once, and if you must have it, take it out of business hours. Politely inform your friends that you must stop, or it will ruin you. If you drink with one you must drink with another, and the opportunity comes too often. When you have finished some difficult work you are to be treated; when you trust you are to be treated; when you accommodate one before another you are to be treated; when you order the stock from the traveling man you are to be treated. Some smiths keep a bottle in a corner to draw customers by; others tap a keg of beer every Saturday for the same purpose. No smith will ever gain anything by this bad practice. He will only get undesirable customers, and strictly temperance people will shun him for it. What he gains on one side he will lose on another. Besides this he will in the long run ruin himself physically and financially. Let the old smith quit and the apprentice never begin this dangerous habit. A smith that is drunk or half drunk cannot do his duty to his customers, and **they know it**, and prefer to **patronize a sober smith.**

RELIGION

TRUE religion is also an uplifting factor, and must, if accepted, elevate the man. I cannot too strongly emphasize this truth. Every smith should connect himself with some branch of the church and be punctual in attendance to the same. There is a great deal of difference between families that enjoy the Christianizing, civilizing and uplifting influence of the church and those outside of these influences. The smith outside of the church, or he who is not a member thereof will, in many cases, be found on Sundays in his shop or loafing about in his everyday clothes, his wife and children very much like him. The church member— his wife and children, are different. Sunday is a great day to them. The smith puts on his best clothes, wife and children the same. Everything in and about the house has a holiday appearance and the effect on them of good music and singing, eloquent preaching, and the meeting of friends is manifested in their language, in their lofty aims, and benevolent acts. Sunday is rest and strength to them.

Brother smiths, six days a week are enough for work. Keep the Sabbath and you will live longer and better.

INCOMPETENCY

Another reason the smith of to-day is not respected is his incompetency.

When a young man has worked a few months in a shop, he will succeed in welding a toe calk on a horseshoe that sometimes will stay, and at once he begins to think he knows it all. There will always be some fool ready to flatter him, and the young man believes that he is now competent to start on his own hook. The result is, he hangs out his shingle, begins to practice horse-shoeing and general blacksmithing, and he knows nothing about either. Let me state here that horse-shoeing is a trade by itself, and so is blacksmithing. In the large cities there are blacksmiths who know nothing about horse-shoeing, as well as horse-shoers who know nothing about blacksmithing, except welding on toe calks, and in many instances even that is very poorly done. In small places it is different. There the blacksmith is both blacksmith and horse-shoer. Sometimes you will find a blacksmith that is a good horse-shoer, but you will never find a horse-shoer that is a good blacksmith. This is not generally understood. To many blacksmithing seems to mean only horse-shoeing, and our trade journals are not much better posted. Whenever a blacksmith is alluded to, or pictured you will always find a horse-shoe in connection with it. Yet there are thousands of blacksmiths that never made a horse-shoe in all their lives. Horse-shoeing has developed to be quite a trade, and if a man can learn it in a few years he will do well. I would not advise any young man to start out for himself with less than three or four years' experience. Every horse-shoer should make an effort to learn blacksmithing. He will be expected to know it, people don't know the difference; besides this, it will, in smaller cities, be hard to succeed with horse-shoeing alone. On the other hand, every blacksmith should learn horse-shoeing, for the same reasons. Therefore, seven or even ten years is a short time to learn it in. But, who has patience and good sense enough to persevere for such a course, in our times, when everybody wants to get to the front at once? Let every young man remember that the reputation you get in the start will stick to you. Therefore be careful not to start before you know your business, and the years spent in learning it will not be lost, but a foundation for your success. Remember, that if a thing is not worth being well done it is not worth being done at all. It is better to be a first-class bootblack or chimney sweep, than be a third-class of anything else.

Don't be satisfied by simply being able to do the work so as to pass, let it be first class. Thousands of mechanics are turning out work just as others are doing it, but you should not be satisfied to do the work as others are doing it, but do it right.

A MODERN GUILD

The blacksmiths and horse-shoers have at last put the thinking cap on, for the purpose of bettering their condition. So far nothing has been accomplished, but I am sure it will, in the long run, if they only keep at it. We are now living in the license craze age. From the saloon keeper down to the street peddler, they all howl for license, and unreasonable as it is, thousands of sensible men will cling to it in hopes that it will help.

We are, more or less, one-idea men, with fads and whims. Nations and organizations are just like individuals, ready to fall into a craze and we see it often. It is natural when we consider that nations and organizations are simple *one* man repeated so many times.

Simply look at the hero-worshiping craze went through at the close of the Spanish war. First, Lieutenant Hobson was the idol, and great was he, far off in Cuba. But, coming home, he made himself obnoxious on a tour through the country, and the worshipers were ashamed of their idol, as well as of themselves. Admiral Dewey was the next hero to be idolized, and he, too, was found wanting.

Physicians have their favorite prescriptions, ministers their favorite sermons. Politicians have their tariff and free trade whims, their gold or silver craze. Mechanics have their one ideal way of doing their work. I know horse-shoers that have such faith in bar shoes that they believe it will cure everything from contraction to heaves. Others have such a faith in toe weight that they will guarantee that in a horse shod this way the front quarters will run so fast that they must put wheels under the hind feet to enable them to keep up with the front feet; and in a three-mile race the front quarters will reach the stables in time to feed on a peck of oats before the hind quarters catch up.

In some States there is a union craze. All that these schemes will do is to prepare the legislatures for the legislation that will some day be asked of them. Unions have been organized and the objections are the same. I object to all these schemes because they fall short of their purpose.

Two years ago the horse-shoers of Minnesota asked the legislature to give them a license law. I wrote to a prominent member of the house of representatives and asked him to put his influence against the measure. He did so, with the result that the bill was killed so far as the counties and smaller towns were concerned. Such a law will only provide for an extra tax on the poor smiths and horse-shoers, and his chances of making a living will not be bettered, because no one will be shut out, no matter how incompetent.

TAXATION WILL NEVER RAISE THE STANDARD OF A MECHANIC

It deprives him of the means whereby to raise himself. Such a law will only create offices to grease the machinery for the political party in power.

THE only thing that will ever elevate the standard of workmanship is education, education and nothing but education. Give us a law that will provide for a certain degree of education before a boy is allowed to serve as an apprentice; and that he will not be allowed to start out for himself until he has served the full term, both as an apprentice and journeyman. And if intemperate, no diploma shall be issued to him. I see now that I was right when I opposed this law. The horse-shoers of Minnesota are now kicking and cursing the examining board. The National Convention of horse-shoers which was held in Cincinnati passed resolutions which were ordered transmitted to the governor of Illinois, requesting that the board of examiners now authorized to grant licenses to horse-shoers in that State, be changed, as "The board has failed to accomplish the purpose for which it was instituted—the elevating of the standard of workmanship of horse-shoers of that State." Unions are all right in every place where there is only one smith, let that smith unite with himself to charge a living price for his work and he is all right. Where there are more than one smith unions will only help the dishonest fellow. Such unions live but for a short time and then the smiths knife each other worse than ever.

In hard times (and hard times are now like the poor, "always with us,") a lot of tinkers start in the shoeing and blacksmith business. If they could make a dollar a day in something else they would stay out, but this being impossible, they think it better to try at the anvil. For them to get anything to do without cutting prices is out of the question, and so the cutting business begins, and ends when the regular smith has come down to the tinker's price. To remedy this we must go to the root of the evil. First, political agitation against a system whereby labor is debased.

This is a fact, in spite of all prosperity howling. Whenever there is trouble between labor and capital we will always find the whole machinery of the government ready to protect capital. The laboring men will not even be allowed to meet, but will be dispersed like so many dogs. They are the mob! But the capitalists, they are gentlemen! When the government wants a tailor for instructor in our Indian schools, or a blacksmith for the reservation, they get about

$600.00 per year. But, when a ward-heeler wants office he must have $5,000 per year. What inducement is it, under such conditions, for a young man to learn a trade? Laboring men, wake up!

But, as this will bring us into politics I shall leave this side of the question, for it would do no good. Thomas Jefferson, in the Declaration of Independence said: "Mankind are more disposed to suffer, while the evils are sufferable, than to right themselves by abolishing the forms to which they are accustomed." The laboring people will, in my judgment, suffer quite a while yet. In the meantime let us build up a fraternity on the ruins of the ancient guilds. Between the twelfth and the fifteenth centuries mechanics of all kinds prospered as never before, nor have they done it since. The reason for this was not a high protective tariff, or anything in that line, but simply the fruit of the guilds and the privilege they enjoyed from the state.

What we now need is a modern guild. I anticipate there would be some difficulty in securing the legislation necessary, but we will not ask more than the doctors now have. I cannot now go into detail; that would take more room and time than I can spare in this book.

ONE thing is certain, we have a hard row to hoe, because, this is a government of injunctions, and any law on the statute book is in danger of being declared unconstitutional, according to the biddings of the money power, or the whim of the judges. One tyrant is bad, but many are worse.

I am no prophet, but will judge the future from the past. History will repeat itself, and Christ's teachings will be found true: "A house divided against itself cannot stand."

I will say so much, however, that no man should be allowed to start out for himself before he has served three years as an apprentice and two or three years as a journeyman. This should be proved by a certificate from the master for whom he has worked. This certificate to be sworn to by his master, one uninterested master and himself. No apprentice to be accepted without a certificate from the school superintendent that he has a certain knowledge in language and arithmetic and other branches as may be required. It shall not be enough to have worked a few days each year, but the whole time. With these papers he shall appear before three commissioners, elected by the fraternity and appointed by the governor of the State.

He shall pay not less than ten and not more than twenty-five dollars for his diploma. All complaint shall be submitted to these commissioners, and they shall have full power to act. If a practitioner acts unbecoming, runs down his competitor, charges prices below the price fixed by the fraternity, or defrauds his customers, such shall be reported to the commissioners, and, if they see fit, they can repeal or call in his diploma and he shall not be allowed to practice in the State. These are a few hints on the nature of the modern guild we ought to establish. The fraternity should have a journal edited by one editor on literature and one on mechanics, the editor on mechanics to be a practical blacksmith with not less than fifteen years' experience. The editors are to be elected by the fraternity. This is all possible if we can get the legislation that the doctors have in many States. And why not?

Mechanics of to-day have a vague and abstract idea of what is meant by journeyman and apprenticeship. In Europe there is yet a shadow left of the guilds where these were in existence.

When I learned my trade I worked some time with my father in Sweden, then I went over to Norway and worked as an apprentice in Mathison & Johnson's machine, file and lock factory of Christiania. I was requested to sign a contract for four years. In this contract was set forth the wages I was to receive, and what I was to learn each year. Everything was specified so that there would be no room for misunderstanding. The first two weeks I worked, they simply drilled me. I was given a good file and a piece of iron, this iron I filed square, round, triangle, hexagon and octagon. I wore out files and pieces of iron one after another, the master giving instructions how to stand, hold the file, about the pressure and strokes of same, etc. The same careful instructions were given in blacksmithing. The apprentice was given some work, and he had to forge it out himself, no matter what time it took, nor did it make any difference if the job, when done, was of any use, the apprentice was simply practicing and accustoming himself to the use of tools. Thus the elementary rules were learned in a few weeks, and the apprentice made capable of doing useful service that would repay for the time lost in the start.

LITERATURE

HAVING thoroughly learned the trade, it is important to keep posted in this matter by reading books and trade journals. As far as books are concerned, we have a few treating on horse-shoeing, with both good and bad ideas. As to blacksmithing, this book, "Modern Blacksmithing," is the first in that line, written by a practical blacksmith and horse-shoer.

Our trade journals must be read with discrimination. They are mostly edited by men having no practical experience in the trade, and are therefore not responsible for the articles these papers contain. Many articles are entirely misleading. Blacksmiths having more experience with the pen than the hammer, and anxious to have their names appear in print, write for these journals.

Prize articles are also doing more harm than good, the judges giving the prizes to men with ideas like their own, not being broad-minded enough to consider anything they don't practice themselves, and the result is a premium on old and foolish ideas.

But we should not stop at this. We should read much. Anything, except bloody novels, will help to elevate the man. No smith should think it idle to read and study. "Every kind of knowledge," observes a writer, "comes into play some time or other, not only systematic study, but fragmentary, even the odds and ends, the merest rag-tags of information." Some fact, or experience, and sometimes an anecdote, recur to the mind, by the power of association, just in the right time and place. A carpenter was observed to be very particular and painstaking in repairing an old chair of a magistrate, and when asked why, said: "I want this chair to be easy for me to sit in some time." He lived long enough to sit in it.

Hugh Miller found time while pursuing the trade of a stone mason, not only to read, but to write, cultivating his style till he became one of the most facile and brilliant authors of the day. Elihu Burritt acquired a mastery of eighteen languages and twenty-two dialects, not by rare genius, which he disclaimed, but by improving the bits and fragments of time which he had to spare from his occupation as a blacksmith.

Let it be a practice or a habit, if you will, to buy at least one book every year, and to read the same, once, twice, thrice, or until its contents are indelibly impressed upon your mind. It will come back to your mind and be useful when you expect it the least.

CHAPTER II

NO other mechanic will try to turn out such a variety of work with so few tools as the blacksmith, even when the smith has all the tools to be had, he has few in proportion to the work. There are a class of smiths who will be content with almost nothing. These men can tell all about the different kinds of tobacco; they can tell one kind of beer from another in the first sip, and the smell of the whisky bottle is enough for them to decide the character of the contents, but when it comes to tools which belong to their trade, they are not in it. It ought to be a practice with every smith to add some new tool every year. But if they are approached on the subject they will generally say, "Oh, I can get along without that." With them it is not a question of what they need, but what they can get along without.

Some smiths have the Chinaman's nature (stubborn conservatism) to the extent that they will have nothing new, no matter how superior to their old and inferior tools; what they have been used to is the best.

When the hoof shears were a new thing I ordered a pair and handed them to my horse-shoer, he tried them for a few minutes and then threw them on the floor and said, "Yankee humbug." I picked them up and tried them myself, and it took a few days before I got used to them, but then I found that they were a great improvement over the toe knife. I told my horseshoer to use them and after a while he could not get along without them, but would yet have used his toe knife if it had not been for the fact that he was compelled to use them. If it was not for the conservatism by which we are all infected more or less, we would be far more advanced in everything.

The mechanic that has poor tools will in every case be left behind in competition with the man with good tools in proper shape. There are smiths who will take in all kinds of shows and entertainments within fifty miles, but when it comes to tools, oh, how stingy and saving they are. There is no investment which will bring such a good return as first-class tools do to a mechanic. The old maxim, "A mechanic is known by the tools he uses," is true. Many of the tools used in the shop can be made by the smith. If less time is spent in the stores and saloon there will be more time for making tools.

I shall, in this chapter, give a few pointers how to make some of the tools used. I will not spend any time in explanation about the more intricate tools like drill presses and tools of that kind, because no smith has experience or facilities to make tools of this character that will be worth anything. I shall simply give a few hints on the most common tools used, with illustrations that will be a help to new beginners. Before we go any further let me remind you of the golden rule of the mechanic, "A place for everything and everything in its place." Some shops look like a scrap iron shed, the tools strewn all over, and one-tenth of the time is spent in hunting for them. I shall first say a few words about the shop and give a plan. This plan is not meant to be followed minutely, but is simply a hint in that direction.

THE SHOP

In building a shop care should be taken in making it convenient and healthy. Most of the shops are built with a high floor. This is very inconvenient when machinery of any kind is taken in for repairs, as well as in taking in a team for shoeing. Around the forge there should be a gravel floor. A plank floor is a great nuisance around the anvil. Every piece cut off hot is to be hunted up and picked up or it will set fire to it. I know there will be some objection to this kind of floor but if you once learn how to keep it you will change your mind. To make this floor take sand and clay with fine gravel, mix with coal dust and place a layer where wanted about four inches thick. This floor, when a little old, will be as hard as iron, provided you sprinkle it every night with water. The dust and soot from the shop will, in time, settle in with it and it will be smooth and hard. It will not catch fire; no cracks for small tools or bolts to fall through; it will not crack like cement or brick floors. If your shop is large then make a platform at each end, and a gravel floor in the center, or at one side, as in figure

1. This floor is cool in summer and warm in winter, as there can be no draft. The shop should have plenty of light, skylights if possible. The soot and dust will, in a short time, make the lightest shop dark. The shop should be whitewashed once a year. Have

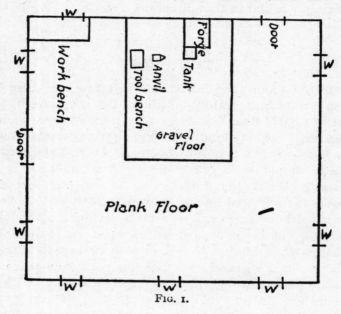

FIG. 1.

plenty of ventilation. Make it one story only if convenient to do so, as an upper story in a blacksmith shop is of very little use. The shop is the place where the smith spends most of his time and he should take just as much care in building it, as a sensible housekeeper does in the construction of her kitchen.

THE FORGE

The forge can be made either single or double, square or round. The square is the best as it can be placed up against the wall, and you will then have more room in front of it. The round forge will take more room, if it is placed in the center of the floor there will be no room of any amount on any side and when the doors are open the wind will blow the fire, cinders and smoke into the face of the smith. This is very uncomfortable. The smokestack, if hung over the fire will sometimes be in the way. Of course the hood can be made in halves and one half swung to the side, but it will sometimes be in the way anyhow, and it seldom has any suction to carry away the smoke and cinders.

THE ANVIL

The anvil should not be too close to the forge, as is often the case in small country shops. Make it six feet from center of fire to center of anvil. The anvil should not be placed on a butcher block with the tools on, but on a timber the same size as the foot of the anvil. Set the timber down in the ground at least three feet. For heavy work the anvil should stand low in order to be able to come down on it with both hammer and sledge with force. When the smith has his hands closed the knuckles of his fingers should touch the face of the anvil and it will be the right height for all-around blacksmithing.

COAL BOX

Close to the forge under the water tank or barrel should be a coal box 18 x 24 x 16 inches, this box to be dug down in the ground and so placed that one end will protrude from under the barrel or tank far enough to let a shovel in. This opening can be closed with a lid if the tools are liable to fall into it. In this box keep the coal wet. In figure 1 a plan is given from which you can get an idea of a shop and how to place the tools and different articles needed.

TOOL TABLES

On the right hand of the anvil should be a tool bench or tool table 20 x 20, a little lower than the anvil. Outside, on three sides and level with the table, make a railing of $1\frac{1}{4}$ inch iron, about $1\frac{1}{2}$ inch space between the table and railing, this makes a handy place for tools and near by. Many blacksmiths have no other place than the floor for their tools, but there is no more sense in that than it would be for a carpenter to throw his tools down on the floor all around him. There ought to be "a place for every tool and every tool in its place."

TOOL TABLE.

THE HAMMER

When a lawyer or a minister makes his maiden speech he will always be in a great hurry on account of his excitement. The sentences are cut shorter, broken, and the words are sometimes only half pronounced. After a few years' practice he will be more self-possessed and the speech will be changed from unintelligible phrases to logical oratory. When the carpenter's apprentice first begins to use the saw, he will act the same way—be in a great hurry—he will run the saw at the speed of a scroll saw, but only a few inches of stroke; after some instructions and a few year's practice the saw will be run up and down steady and with strokes the whole length of the blade. When the blacksmith's apprentice begins to use the hammer he acts very much the same way. He will press his elbows against his ribs; lift the hammer only a few inches from the anvil and peck away at the speed of a trip hammer. This will, in most cases, be different in a few years. He will drop the bundle—that is, his elbows will part company with his ribs, the hammer will look over his head, there will be full strokes and regular time, every blow as good as a dozen of his first ones. Some smiths have the foolish habit of beating on the anvil empty with the hammer, they will strike a few blows on the iron, then a couple of blind beats on the anvil, and so on. This habit has been imported from Europe, free of duty, and that must be the reason why so many blacksmiths enjoy this luxury.

THE SLEDGE

In Europe great importance is laid upon the position taken by the apprentice and the manner he holds the sledge. The sledge is held so that the end of it will be under his right armpit, when the right hand is next to the sledge, and under his left arm when the left hand is nearer the sledge. In this unnatural position it is next to impossible to strike hard and do it for any time. This is another article imported free of duty, but few Americans' have been foolish enough to use it. In this country the apprentice will be taught to use the tools in a proper way.

The end of the sledge-handle will be to one side; at the left, if the left hand is at the end of the handle, and at the right if the right hand is at the end of the handle; and be down between his feet when the handle's end must be low. The apprentice should stand directly in front of the anvil.

In swinging, the sledge should describe a circle from the anvil close down to the helper's feet and up over his head and down to the anvil; this is a perpendicular circle blow. Be sure not to give it a horizontal start; that is, with one hand close to the sledge the apprentice starts out either in the direction of the horn or the butt end of the anvil, and then up while both hands should clasp the extreme end of the handle close together the sledge should be dropped down to the feet then up. The hold taken should not be changed, but the hands held in the same place. (See figure 4.)

For ordinary use a nine-pound sledge is heavy enough, a large sledge will give a bump, while a small one will give a quick good blow; it is only occasionally and for special purposes a large sledge is needed, even an eight-pound sledge will do. Try it, and you will be surprised how nice it works.

With these preliminary remarks we shall now begin to make a few tools. We will begin with the blacksmith's tongs. I shall only give an idea how to forge the jaws, and every man that needs to make them has

FIG. 4.

seen enough of this simple tool to know what kind is needed, and what he has not seen will suggest itself to every sensible smith.

BLACKSMITH'S TONGS

Take a piece of one-inch square Swede iron, **hold the iron** diagonally over the anvil, with your left hand a little toward the horn, the end of the iron to reach out over the outside edge of the anvil. Now strike so that the sledge and hammer will hit half face over the anvil and the other half of the sledge and hammer out-

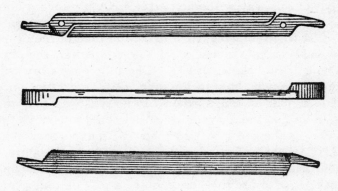

side of the anvil. Hammer it down to about three-eighths of an inch thick. Now pull the iron towards you straight across the anvil, give it one half turn toward yourself so that this side which was up, now will be towards yourself; the end that first was outside the anvil now to rest over the inner edge of the anvil, push the jaw up against the anvil until it rests against the shoulder made in the first move. Now hammer this down until it is the thickness of the jaw that is desired. Next, turn it over, with the bottom side up or the side that was down, up; push it out over the outside edge of the anvil again so far that the shoulder or set down you now have up, will be about an inch outside and over the edge of the anvil, now give a few blows to finish the jaw, then finish the shanks and weld in half inch round iron to the length desired. The jaws should be grooved with a fuller, if you have none of the size required take a piece of round iron and hammer it down in the jaws to make the groove. Tongs grooved this way will grip better. Next, punch a hole in one jaw, place it over the other in the position wanted when finished, then mark the hole in the other jaw, and when punched rivet them together, the jaws to be cold and the rivet hot. The following story will suggest to you how to finish it. An apprentice once made a pair of tongs when his master was out, and when he had them riveted together could not move the jaws. As he did not know how to make them work he laid them away under the bellows. At the supper table the apprentice told his master the following story: An apprentice once made a pair of tongs and when he had them riveted together he could not move the jaws, and as he did not know what to do he simply threw them away, thinking he must have made a mistake somehow. "What a fool," said the master, "Why didn't he heat them." At the next opportunity the apprentice put his tongs in the fire and when hot they could be worked very easily.

HOW TO MAKE A HAMMER

Take a piece of tool steel $1\tfrac{1}{4}$ inches square, neat it red hot. Now remember here it is that the trouble begins in handling tool steel. If, in the process, you ever get it more than red hot, it is spoiled, and no receipt, or handling or hammering will ever make it good again. The best thing in such a case is to cut off the burnt part in spite of all proposed cures. This must be remembered whenever you heat tool or spring steel. If the burnt part cannot be cut off, heat it to a low heat, cool it in lukewarm water half a dozen times, this will improve it some, if you can hammer it some do so. Now punch a hole about two inches from the end with a punch that will make a hole $1\tfrac{1}{8}$ x $\tfrac{3}{8}$. If the punch sticks in the hole, cool it off and put a little coal in the hole that will prevent the punch from sticking. This is a good thing to do whenever a deep hole is to be punched. Be sure that the hole is made true. Next, have a punch the exact size of the hole wanted when finished, drive it in and hammer the eye out until it has the thickness of about $\tfrac{3}{8}$ of an inch on each side and has a circle form like No. 2, Figure 5.

In order to do this you may have to heat the eye many times, and upset over it with the punch in the eye. This done put in the bottom fuller and with the top fuller groove it down on each side of the eye, like the cut referred to. Now dress down the face then the peen-end. When finished harden it in this way: Heat the face-end first to a low red heat, dip in water about an inch and a half, brighten the face and watch for the color. When it begins to turn blue cool off but don't harden the eye. Wind a wet rag around the face end and heat the peen-end, temper the same way. With a piece of iron in the eye, both ends can be hardened at the same time, but this is more difficult, and I would not recommend it.

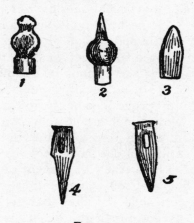

Fig. 5.

For ordinary blacksmithing a flat peen hammer is the thing, but I have seen good blacksmiths hang on to the machinist's hammer as the only thing. See No. 1, Figure 5. This hammer is more ornamental than useful in a blacksmith shop. The hammer should be of different sizes for different work, light for light work, and for drawing out plowshares alone the hammer should be heavy.

For an ordinary smith a hammer of two up to two and one-half pounds is right. Riveting hammers should be only one pound and less. No smith should ever use a hammer like No. 3, in Figure 5. This hammer I have not yet been able to find out what it is good for. Too short, too clumsy, too much friction in the air. I have christened it, and if you want my name for it call it Cain's hammer. It must surely look like the hammer used by him, if he had any.

HOW TO MAKE CHISELS

A chisel for hot cutting, see Figure 5, No. 4. This chisel is made of $1\frac{1}{4}$ square tool steel. Punch a hole $1\frac{1}{8}$ x $\frac{1}{4}$ x $\frac{1}{2}$ about three inches from the end, the eye should be narrow in order to leave material enough on the sides to give it strength. When eye is finished, forge down below it, not on the head-end, with top and bottom fullers, like cut. This gives the chisel a better shape. Now dress down the edge, then heat to a low cherry red, and harden, brighten it and when the color is brown cool off.

COLD CHISELS

Use same sized steel as above referred to, make it like No. 5, Figure 5. To distinguish it from the hot cutting chisel, and to give it more strength, in hardening this chisel, draw the temper until it is blue. This is the right temper for all kinds of cold chisels.

SET HAMMER

One might think that anybody knows how to make a set hammer, if every smith knows it, I don't know, but I do know that there are thousands of smiths who have never had a set hammer nor know its use. To make one: Take a piece of tool steel 1¼ x 1¼ inches, punch a hole about two inches from the end, the hole to be 1¼ x ⅜. Now cut off enough for head. Make the face perfectly square and level, with sharp corners, harden and cool off when the temper turns from brown to blue. This is a very important little tool and for cutting steel it is a good deal better than the chisel. Plow steel of every kind is easier cut with this hammer than any other way. In cutting with the set hammer hold the steel so that your inner side of the set hammer will be over the outside edge of the anvil. Let the helper strike on the outside corner of the set hammer and it will cut easy. The steel to be cut should be just a little hot, not enough to be noticed. If the steel is red hot the set hammer cannot cut it. The heat must be what is called blue heat. I would not be without the set hammer for money, and still I often meet smiths who have never seen this use made of the set hammer. Plow points, corn shovels, and seeder shovels are quicker cut with this tool than any other way, with the exception of shears.

TWIST DRILLS

Twist drills are not easy to make by hand, as they should be turned to be true, but a twist drill can be made this way. Take a piece of tool steel round and the size of the chuck hole in your drill press. Flatten it down to the size wanted, heat, put the shank in the vise, take with the tongs over the end and give one turn to the whole length, turn to the left. When finished be sure that it is not thicker up than it is at point, and straight. Now harden, heat to a low cherry red, cool off in luke-warm water—salt water, if you have it—brighten it and hold over a hot bar of iron to draw temper, cool off when brown, the whole length of the twist should be tempered.

Another way to make a drill is to just flatten the steel and shape to a diamond point and bend the shares forward. This is a simple but good idea and such drills cut easy. In cooling for hardening turn the drill in the water so that the edge or shares are cooled in proportion to point, or the shares will be too soft and the point of such a drill too hard. Our trade journals, in giving receipts for hardening drills, often get watch-makers receipts. This is misleading: watch-makers heat their drills to a white heat. Now, remember, as I have already said, when your drill or tool of this kind is heated to this heat the best thing to do is to cut that part off. It is different with watchmakers, they do not look for strength, but hardness. They run their drills with a high speed, cut chips that cannot be discerned with the naked eye, and must have a drill that is hard like a diamond. For drilling iron or steel the drill does not need to be so very hard, but tough rather, because of the slow speed and thick chips. Few smiths have been able to master the simplest tempering, and they think if they could get a complicated receipt they would be all right. We are all more or less built that way. Anything we do not

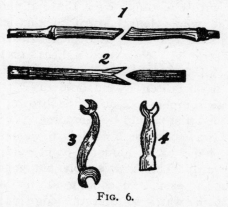

Fig. 6.

understand we admire. Simple soft water and the right heat is, in most cases, the only thing needed for hardening. I had occasion to consult a doctor once who was noted for his simple remedies. A lady got some medicine and she wanted to know what it was so she could get it when the doctor was not at home, but he refused to reveal it to her. When the lady had left the doctor told me the reason why. "This lady," said the doctor, "does not believe in simple remedies which she knows, but believes in those remedies she knows nothing about." I think it is better for us to try to understand things and not believe much in them before we understand them.

S WRENCH

See Figure 6, No. 3. This wrench is for ⅜ nut on one end and ½ on the other, just the kind for plow work. To make one, take a piece of tool steel 1½ x ⅝, start as you see in No. 4, Figure 6. Set the jaws down with the fullers, punch a round hole as in end No. 4, cut out from hole and finish the jaws to make the right length, now bend it in S shape and finish. This makes the best wrench. Do not heat over a red heat.

ROCK DRILLS

Few blacksmiths know how to make a rock drill. Take a piece of round or octagon steel, the desired length and thickness, shape it, but it must be remembered that if during the process you ever get it over a red heat there is no use to proceed, but just cut off that much and start again, no hardening will prevail if it is burnt. The trouble begins when you put the steel into the fire, and you must watch until you have it finished. When ready to harden heat it to a cherry red heat, cool in water not too cold, brighten and watch for temper. When it is yellow, cool it off, but not entirely, take it out of the water before it is quite cold and let it cool slowly, this will make the drill both hard and tough. By this simple process I have been able to dress drills and get such a good temper than only two per cent would break. Another way to harden is to heat to a very low heat and cool it off entirely at once. A third way is to temper as first stated and when yellow set the drill in water only one half an inch deep and let it cool. By this process a good per cent will break just at the water line.

CHAPTER III

HOW TO STRIKE AND TURN THE IRON— RULES FOR SMITH AND HELPER

HE smith should never turn the iron on the helper's blow, he should turn on his own blow, that is, never turn the iron so that the helper's blow will hit it first because he is not prepared for it and cannot strike with confidence, but the smith will not be bothered by turning the iron for himself as he knows when he turns and is prepared for it. The smith should strike the first blow in starting, or signal the helper where to strike, in case the smith cannot strike the first blow. The smith calls the helper by three blows on the anvil with his hammer, and when the smith wants the helper to cease striking he taps with the hammer twice on the anvil. The helper should strike the blow he has started when the smith signals him to stop. The helper should watch the time of the smith's hammer; if fast, keep time with it, if slow, keep time with it. The helper should strike where the smith strikes or over the center of the anvil. The helper should always lift the sledge high, in order to give the smith a chance to get in with the hammer.

THE FIRE

It is proper before we go any farther to say a few words about the fire.

An old foreman in the blacksmith department of a factory told me once in a conversation we had about

"CORRECT POSITION" AT THE ANVIL

the fire, that he had come to the conclusion that very few blacksmiths have learned how to make a good fire. It takes years of study and practice before the eye is able to discern a good fire from a bad one. A good fire must be a clear fire, the flame must be concentrated and of a white color. Even the nose must serve to decide a bad fire from a good one. A strong sulphur smell indicates a poor fire for welding. In order to get a good fire there must be, first, good coal; second, plenty of it. It is no use to pile a lot of coal on an old fire, full of cinders and slag. The fire-pot must be clean. Many blacksmiths are too saving about the coal. They take a shovel of coal, drop it on the forge in the vicinity of the fire and sprinkle a handful of it in the fire once in a while. In such a case it is impossible to do good work and turn it out quick. Have a scoop shovel and put on one or two shovels at a time, the coal should be wet. Then pack it in the fire as hard together as you can. Sprinkle the fire with water when it begins to spread. In this way you get a hard fire. The flames are concentrated and give great heat. Saving coal is just like saving feed to a horse, or grub to your apprentice. Neither will give you a good day's work unless he has all he wants to eat. The fire, of course, should be in proportion to the work, but in every case should the fire be large enough to raise it up from the tuyer iron as much as possible. In a small fire the blast strikes directly on the iron and it begins to scale off; in a good fire these scales melt and make it sticky, while in a low and poor fire the scales blacken and fall off. This never happens if the fire is full of good coal and high up from the tuyer iron.

Good strong blast is also necessary for heavy work. There is an old whim about the fire that everybody, farmers and others, as well as blacksmiths, are infected with, and that is, if a piece of brass is put in the fire it renders the fire useless to weld with. Now, while it is a fact that brass is not conducive to welding it takes a good deal of it before the fire is made useless. One smith will not dare to heat a galvanized pipe in his fire, for fear it will spoil it, while another smith will weld a piece of iron or steel to such a pipe without difficulty. Don't swear and curse if the fire is not what you expect it to be, but simply make it right. Some smiths have the habit of continually poking in the fire, if they weld a piece of iron they never give it rest enough to get hot, but turn it over from one side to another and try to fish up all the cinders and dust to be found in the fire. This is a bad habit. Yellow colored fire is a sign of sulphur in the fire and makes a poor fire for welding. Dead coal makes a poor fire.

TUYER IRON

One of the chief reasons for a poor fire is a poor blast. No patent tuyer will give blast enough unless you run it by steam and have a fan blower. Ninety per cent of the blast is lost in transmission through patent tuyers. The only way to get a good blast is to have a direct tuyer, and one with a water space in.

To make a direct tuyer take a pipe $1\frac{1}{4}$ x 12 inches long, weld around one end of this pipe an iron $3\frac{5}{8}$ to make it thick on the end that is in the fire, flare out the other end for the wind pipe to go in and place it horzontal in the fire and fill up around it with fireproof clay. This gives the best fire. The only objection to this tuyer is that where soft coal is used, as is mostly the case in country shops, it gets hot and clogs up, but with a strong blast and good hard coal it never gets hot, provided the fire is deep enough. From five to eight inches is the right distance from the tuyer to the face of the fire. In factories this kind of tuyer is used, and I have seen them used for ten years, and

WATER TUYER

never found them to clog once. The tuyer was just as good after ten years use as it was when put in.

To make a water tuyer take a pipe $1\frac{1}{4}$ x 12, weld a flange on each end for water space, now weld another pipe over this, and bore holes for $\frac{1}{4}$ inch pipes in the end, where the blast goes in. One hole on the lower or bottom side should be for the cold water to go in through, and one hole on the upper side for the hot water to go out through. These pipes to connect with a little water tank for this purpose. The pipes should be watched so that they will not be allowed to freeze or clog, as an explosion might follow. These tuyers never clog. I now use one that I have made as above described. The dealers now have them to sell. Any smith can get them as they are hard to make by the average smith.

BLOWERS

I have tried many kinds of blowers and I shall give my brother smiths the advantage of my experience.

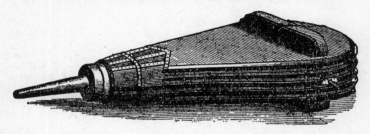

Portable forges run with fan blowers are fair blowers if you are strong enough to pump away at high speed, but it takes a horse to do that, and as soon as you drop the lever the blast ceases. Root's blower works easier, but the objection is the same, as soon as you drop the crank the blast stops. Besides this trouble, this blower is often in the way. I have never found anything to beat the bellows yet, if you only know how to use them.

Never take a set of bellows less than 48 extra long. Cut the snout off so that it will give a hole 1½, and with a water tuyer this blower cannot be beaten, except by a fan blower run by steam. The bellows should be hung over head to be out of the way. When these bellows are full of wind they will blow long enough after you have dropped the lever to do quite a good many things around the forge, and to handle the iron in the fire with both hands as is often necessary.

WELDING IRON

Welding iron is easy and no other welding compound is needed than sand, unless it is a case when the iron is liable to burn or scale off, borax will prevent this. There are three kinds of welds, butt, lap and split. The butt weld is most used in welding iron. The ends should be rounded off a little so that the center will weld first. Weld the ends this way either in the fire or on the anvil, butting the ends while you strike over and dress down the weld. In welding lap welds upset the ends and make them a good deal heavier than the size of the iron is; then lap the ends with a short lap. New beginners will always make a long lap. This is wrong, for if the lap is long it will reach beyond the upset part and the ends cannot then be welded down, without you make it weak. If soft steel is welded cut a short cut with the chisel in the center of the lap, as shown in Figure 6, No. 1. This cut will hook and prevent the ends from slipping; if properly prepared this weld will not show at all when done.

SPLIT WELDS

Split weld is preferable when steel is to be welded, especially tool steel of a heavy nature, like drill bits for well drillers.

If the steel is welded to iron, split the iron and draw out the ends as thin as possible and make it the shape shown in Figure 6, No. 2. Taper the steel to fill the split made in the iron, when it fits perfectly cut beard in it to catch in the lips of the iron when fitted in. See Figure 6, No. 2. When finished heat the split end and cool off the tapered end. Place the tapered end snug up in the split and hammer it together with a heavy sledge. If there is any crack or opening at the end of the tapered end, plug it up with iron plugs, if this is not done, these holes will be almost as they are, because it is hard to weld a heavy shaft or drill, or rather, it is hard to hammer them together so the holes will close in. Now heat, but if you have tool steel go slow, or your steel will burn before the iron is hot enough. Weld the lips while the rod or drill is in the fire. For this purpose use a hammer with an iron handle in. When the lips are welded all around take it out and let two good helpers come down on it with all their might. When welded smooth it up with the hammer or flat hammer.

WELDING STEEL

Welding steel is quite a trick, especially tool or spring steel. The most important part to remember is, to have a good clean fire, and not to over heat the steel. To a good smith no other compound is needed than borax, but if this is not sufficient, take some borings from your drill, especially fine steel borings, and cover the weld with this and borax, and if a smith cannot weld with this compound there is no use for him to try. Most of the welding compounds are inferior to this, but some smiths would rather believe in something they don't know anything about; another will not believe in anything he can get for nothing.

BANDS OR HOOPS

When a round object is to be ironed or a hoop put on to anything round, measure, that is, take the diameter then multiply by three, add three times the thickness of the iron (not the width), add to this one time the thickness of the hoop for the weld and you have the exact length of the iron needed; in other words, three times the diameter, four times the thickness of the band. This is a simple rule, but I know a good many old smiths who never knew it.

SEEDER SHOVELS

To weld seeder shovels is no easy job. Prepare the shovel; shape almost to it proper shape, draw out the shanks, weld the points first, heat shovel and shank slow, then fit them together so that no cinders can get in between. Now remember, if your fire is not at least five inches up from the tuyer iron, and clear, it is no use to try. Hold your shovel in the fire, shank down. Heat slow, use borax freely and apply it on the face side of the shovel to prevent it from burning. When ready, weld it over the mandrill and the shovel will have the right shape. If soft center, harden like a plow lay.

DRILLING IRON

Every smith knows how to drill, sometimes it gives even an old smith trouble. The drill must be true, the center to be right, if one side of the drill is wider than the other or the drill not in proper shape the hole will not be true. For centuries oil has been used for drilling and millions of dollars have been spent in vain. It is a wonder how people will learn to use the wrong thing. I don't think that I have ever met a man yet who did not know that oil was used in drilling. In drilling hard steel, turpentine or kerosene is used as oil will then prevent cutting entirely. Nothing is better than water, but turpentine or kerosene is not as bad as oil; if you think water is too cheap use turpentine or kerosene. I had occasion once to do a little work for a man eighty years old, and when I drilled a hole, used water. The old man asked if water was as good as oil, and when informed that it was better, said: "I used to be quite a blacksmith myself, I am now eighty years old, too old to do anything, but I am not too old to learn." It ought to suggest itself to every smith that while oil is used in boxes to prevent cutting, it will also prevent cutting in drilling.

HOW TO DRILL CHILLED IRON

First prepare a drill which is thicker at the point than usual, and oval in form, then harden it as follows: heat to a low cherry red heat and cool in the following hardening compound: two quarts soft water, one-half ounce sal-ammoniac, salt, three ounces. Don't draw the temper, for if you have the right heat you will get the right temper. Now drill and use water, not oil. Feed carefully but so the drill will cut right along. If you have no chance to get the compound, harden in water but draw no temper, let it be as hard as it will.

If the iron is too hard to be drilled and you can heat the same do so, heat to a low red heat and place a piece of brimstone just where the hole is to be; this will soften the iron through, so the hole can be drilled. Let it cool slowly.

STANDING COULTERS

Standing coulters are made of different materials and of different shapes. Take a piece of iron $2\frac{1}{4}$ x $\frac{1}{2}$, twenty-eight inches long. Cut off the end after you have thinned it out about 5 inches from the end, cut diagonally Now weld the cut-off piece to the main shank. The cut-off piece to be laid on the outside and welded, bend the iron as soon as it is welded so that it has the shape of the coulter, draw out a good point and sharpen the iron just the same as if it was a finished coulter. This done, cut off a piece of steel, an old plow lay that is not too much worn will do, cut

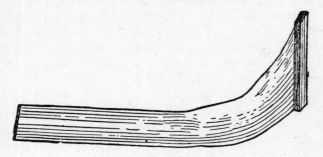

STANDING COULTER

the shape of the coulter you have now in the iron, and let the steel be half an inch wider than the iron, but on the point let it be as long as it will, because the point ought to be quite long, say about nine inches. Next draw the steel out thin on the upper end, heat the iron red hot, place it on the anvil outside up, put a pinch of borax on it at the heel, then a pinch of steel borings, place the steel on top of this and keep in position with a pair of tongs; now hold it on the fire heel down, and heat slow. When it is hot let the helper strike a pressing blow or two on it and it will stick until you have taken the next weld. Put borings, and borax between steel and iron for each weld. When finished, the angle should be that of the square; that is, when you place the coulter in the square the shank should follow one end of the square and the foot of the coulter the other. The edge of the outside side should follow the square from the point up. When it does it looks like a hummock in the coulter but it is not. Old breakers will be particular about this as it will cut a clean furrow if it is made in this way and it will work easier. If the edge stands under the square the coulter will wedge the plow out of land and make a poor furrow. Next finish the chisel point, soft or hard steel as you please; weld it to the coulter on the inside, that is, the side next to the furrow.

Last punch or drill the hole in the heel. The coulter should not be hardened except a little on and along the point. There is no need of a double chisel point, such a point will be too clumsy and run heavy. I have received a premium on a coulter made in this shape.

MILL PICKS

Mill picks are very easily dressed and hardened, the whole trick in this case, as in many others, lies in the right heat of the steel. Be careful not to heat to a higher than a red heat. Dress the pick and temper with a low heat, when the color is dark yellow the temper is right, if the steel is of the right kind. No other hardening compound is necessary than water. After a little experience any smith can do this work first class.

A smith once wanted to buy my receipt for tempering. He believed I had a wonderful prescription, or I could not succeed as I did. I told him I used only water, but he insisted that I was selfish and would not reveal it to him.

If tools and receipts would do the work there would be no need of experienced mechanics. Tools and receipts are both necessary, but it must be a skilled hand to apply them.

HARDEN FILES

The best way to harden files is to have a cast iron bucket filled with lead. Heat it until the lead is red hot, then plunge the file into this, handle up. This will give a uniform heat and the file will not warp so easy if the heat is right. In cooling the file off, use a box four or five feet long with salt water in, run the file back and forth endwise, not sideways, that will warp the file, take it out of the water while yet sizzling. Now, if warped, set it between a device so that you can bend it right. While in this position sprinkle water over where you straighten until cold and the file will be right.

HARDEN TAPS AND DIES

Heat the tap or die to a red cherry, cool off entirely in water, brighten with an emery paper. Now, hold over a hot iron until the tap or die has a dark straw color, then cool off. If a light tap, the temper can be drawn over a gaslight, using a blowpipe.

BUTCHER KNIFE

To make a butcher knife, one smith will simply take an old file, shape it into a knife, and harden. The best way to make a knife is to first draw out a piece of iron ¾ inch wide and $\frac{1}{8}$ of an inch thick, twice the length of the knife. Prepare the steel the same width as the iron, ⅛ of an inch thick, weld this steel in between the iron. This will make a knife that will not break. When ready to harden heat to a low red heat, cool off entirely in water. Brighten and hold over a hot iron until brown, then cool off.

The steel should be good tool steel, a flat file will do, but the cuts must be ground or filed off entirely before you touch it with the hammer, for if the cuts are hammered in they will make cracks in the edge of the knife, and the same will break out.

HOW TO REPAIR CRACKED CIRCULAR SAWS

If a circular saw is cracked it can be repaired so that the crack will go no further, and if the crack is deep, it can be so remedied that there will be no danger in using it. Ascertain the end of the crack, then drill a $\frac{3}{16}$-inch hole so that the crack will end in that hole. Countersink on each side and put in a rivet. Don't let the rivet stick its head over the face of the saw.

If the crack is deep put another rivet about half an inch from the edge. If the saw is too hard to drill, heat two irons about 1 square or round, square up the ends and set the saw between the ends so that they will meet over the place where the hole is to be drilled. When the saw is dark blue, the temper is out. It might be a possibility that this will spring the saw in some cases, therefore, I advise you to try drilling the hole without any change in temper. Prepare a drill that is harder than usual, use no oil, but water.

HOW TO PREVENT A CIRCULAR SAW FROM CRACKING

The reason why a circular saw cracks is, in most cases, incorrect filing. In filing a saw, never let a flat file with its square corners touch the bottom of the teeth you are filing; if you do, you will make a short cut that will start the crack. The best way is to gum the saw in a saw gummer or on an emery wheel, or use a round-edged file.

HOW TO SEW A BELT

Belts can be riveted, sewed, or hooked together. A new leather belt should not be riveted, because such a belt will stretch and have to be cut out and sewed over quite often at first. There are hooks made of steel for belt sewing, these are all right when the pulleys are not less than six inches in diameter and the speed is slow. In using the hooks be careful not to

MODERN BLACKSMITHING

bend them too sharp or drive the bends together too hard; in so doing they will cut through the leather and pull out. Lacing is the best for all kinds of belts.

In sewing a belt with lacing, first punch with a punch made for this purpose, holes in proportion to the width. Don't punch them too close to the ends. Begin sewing in the center holes and start so that both ends of the lacing will come out on the outside of the belt. Now sew with one end to each side, and be careful not to cross the lacing on the side next to the pulleys. The lacing should be straight on that side. When the belt is sewed punch a small hole a little up in the belt to receive the last end of the lacing; the last end should come out on the outside of the belt. In this end cut a little notch about three-fourths through the lacing close to the belt, and then cut the lacing off a quarter of an inch outside of this notch. This notch will act as a prong and prevent the lacing from pulling out. Tap it lightly with a hammer above the seam to smooth it down.

POINTS ON BELTS

In placing shafts to be connected by belts, care should be taken to get the right working distance one from the other. For smaller belts 12 to 15 feet is about the right distance. For large belts, a greater distance is wanted. The reason for this is that when pulleys are too close together there is no sag in the belts and they must therefore be very tight in order to work.

Belts should not have too much sag, or they will, if the distance between the pulleys is too far apart, produce a great sag and a jerking motion which will be hard on the bearings. Never place one shaft directly over another, for then the belts must be very tight to do the work, and a tight belt will wear out quicker and break oftener in the lacing than a loose one; besides this the bearings will give out sooner.

If a belt slips use belt oil or resin, or both.

BOB SHOES

In repairing old bob sleds is is difficult to find shoes to suit. But in every case the shoe can be fitted to suit without touching the runner. The trick here as in many other cases in the blacksmith business, lies in the heating. Any shoe can be straightened or bent to fit the runner if only heated right. A low cherry-red heat and a piece if iron to reach from the crooked end of the shoe and far enough back to leave a space between where it wants to be straightened. Now put it in the vise and turn the screws slowly and the shoe will stand a great deal. If too straight, put the shoe in between a couple of beams so that you can bend it back to the right shape. Remember the heat.

I have put on hundreds and never knew of a shoe that broke when the heat was right. I must confess, however, that my two first shoes broke, but I think I learned it cheap when I consider my success after that. The shoe should fit the runner snug. Ironing bobs is a very simple and easy thing, every blacksmith, and even farmers sometimes, are able to iron their own sleds fairly well, and I don't think it will be of much interest for the readers of this book to treat that subject any further.

AXES AND HATCHETS

Dressing axes is quite a trick and few blacksmiths have mastered it. It is comparatively easy when one knows how. I have several times already warned against over heating and if this has been necessary before, it is more so now in this case. In heating an ax do not let the edge rest in the center of the fire, it will then be too hot at the edge before it is hot enough to hammer it out. Place the edge far enough in to let it over the hottest place in the fire. Go slow. When hot, draw it to the shape of a new axe, don't hammer on one side only. In so doing the ax will be flat on one side and curved up on the other. If uneven trim it off; trim the sides also if too wide; don't heat it over the eye; be sure you have it straight. When ready to harden, heat to a low red heat and harden in luke warm water. The heat should be only brown if it is a bright sunny day. Brighten and look for the temper. You will notice that the temper runs uneven;

it goes out to the corners first, therefore dip them (the corners) deeper when cooling, and with a wet rag touch the place on the edge where the temper wants to run out. Some smiths, when hardening, will smear the ax with tallow instead of brightening it, and hold it over the fire until the tallow catches fire, then cool it off. This is guess work, and the axe is soft in one place and too hard in another. The best way is to brighten the ax and you can see the temper, then there is no guess work about it. When blue cool it partly off and then while the ax is still wet you will observe under the water or through the water a copper color. This color will turn blue as soon as the ax is dry, and is the right color and temper. Cool it slowly, don't cool it off at once, but let it cool gradually, and it will be both hard and tough.

By this simple method I have been very successful, breaking only three per cent, while no new ax of any make will ever do better than ten per cent. Some will even break at the rate of twelve and thirteen per cent.

The ax factories, with all their skill and hardening compounds, have to do better yet to compete with me and my simple method.

WELL DRILLS

Well drills are made of different sizes and kinds. Club bits and Z bits. How to dress: heat to a low red heat. If nicked or broken, cut out, otherwise draw it out to the size wanted. The caliper should touch the lips of the bit when measured diagonally so that the bit has the size on all corners. Heat to a low red heat and harden, the temper to be from dark straw color to blue according to the kind of drilling to be done. The trick, in two words, low heat.

GRANITE TOOLS

By granite tools is meant tools or chisels used by granite or marble workers for cutting inscriptions on tombstones.

When a man understands how these tools are used it is easier to prepare them. These are the kind of tools where an unusual hardness is required. The hammer used in cutting with this chisel is very small, and the blow would not hurt your nose, so light it is, therefore they will stand a high heat and temper. The chisels should be very thin for this work. When dressed and ready to harden, heat to a red heat and harden in the following solution: one gallon soft water, four ounces salt. Draw the temper to a straw color.

A blacksmith once paid a high price for a receipt for hardening granite tools. The receipt was, aqua, one gallon; chloride of sodium, four ounces. This receipt he kept as a secret and the prepared compound he bought at the drug store, thus paying 50 cents for one gallon of water and four ounces of salt. The real worth is less than a cent. It is said he succeeded remarkably well with his great compound, which he kept in a jug and only used when anything like granite tools were to be hardened. The reason why he succeeded so well was because of his ignorance concerning his compound, not because it was not good enough. I hold that it is one of the best compounds, in fact, the best he could get. People in general like to be humbugged. If they only get something new or something they don't know anything about, then they think it wonderful.

Salt and water should be called salt and water, and be just as much valued. Let us "call a spade a spade," the spade will not be more useful by another name, nor will it be less useful by calling it by its proper name.

CHAPTER IV

HEN vehicles were first used is hard to tell, but we know that they have been used for thousands of years before the Christian era. It is easy to imagine how they looked at that time, when we know how half-civilized people now make wagons. The first vehicle was only a two-wheeled cart called chariot. Such chariots were used in war and that it was a case of "great cry and little wool" is certain.

The blacksmith used to be the wagon and carriage maker. Now it is only a rare case when a blacksmith makes a carriage, and when it happens most of the parts are bought. In 1565 the first coach was made in England.

Now there are hundreds of factories making wagons and carriages and parts of them for repair use by blacksmiths and wagon makers. It is no use for any blacksmith or wagon maker to compete with these factories. We have neither the means nor the facilities to do it, and have to be content with the repairs they need. The most important repairs are the setting of tire, welding and setting axle stubs.

SETTING TIRE

Wagon tire is often set so that more harm than good is done to the wheel.

In setting tire the first thing to do is to mark the tire. Many blacksmiths set tires without marking the tire. This is poor work. In order to do a good job the tire should be set so that it is in the same place it had. There are generally some uneven places in the fellows and when the tire is set the first time, it is hot all around and will settle down in these low places. Now, if the tire is not marked and set back in its exact bed, it will soon work loose again, and it is liable to dish the wheel too much as it don't sink into its place, but is held up in some places. Another thing, when a tire is worn so that it becomes thin it will settle down on the outside, especially when the wheel is much dished. Now if you reverse the tire it will only touch the fellow on the inner edge of the wheel, and leave an open space between the fellow and the tire on the outside. When a wheel has bolts every smith knows that it will make trouble for him if he don't get the tire back where it was. In every case take a file or a chisel and cut a mark in the tire near to the fellow plates, cut also a light mark in the fellow. These marks are to be on the inside of the wheel: 1, because it will not be seen on that side; 2, because in putting the tire on, the wheel should be placed with that side up. If there are nails in the tire cut them off with a thin chisel so that it will not mark the fellow, or drive them into the fellow with a punch. Next, measure the wheel with the gauge (the wheel is supposed to be right, not fellow bound nor any spokes loose in the tenon). This done, heat the tire and shrink it. If the wheel is straight give it half an inch draw, sometimes even five-eighths if the wheel is heavy and strong. But if the wheel is poor and dished, do not give it more than one-fourth-inch draw. One tire only with a little draw can be heated in the forge, but if there is more than one tire heat them outside in a fire made for this purpose, or in a tire heater.

There are different ways of cooling the tire. Some smiths have a table in a tank, they place the wheel on the table and with a lever sink both wheel and tire in the water. There are many objections to this. 1, You will have to soak the whole wheel; 2, it is inconvenient to put the tire on; 3, in order to set the tire right, it is necessary to reach the tire from both sides with the hammer; 4, when spokes have a tendency to creep out, or when the wheel is much dished, the wheel should be tapped with the hammer over the spokes. Now, to be able to perform all these moves, one must have, first, a table; this table to be about twelve inches high and wide enough to take any wheel, with a hole in the center of table to receive the hub. On one side you may make a hook that will fall over the wheel and hold the tire down while you get it on. Close to this table have a box 5½ feet long, 12 inches wide and 12 inches deep. On each side bolt a piece of two by six about three feet long. In these planks cut notches in which you place an iron rod, run through the hub. On this rod the wheel will hang. The notches can be made so that any sized wheel will just hang down enough to cover the tire in the water. In this concern you can give the wheel a whirl and it will turn so swift that there will be water all around the tire. It can be stopped at any time and the tire set right, or the spokes tapped. With these accommodations and four helpers I have set six hundred hay rake wheels in nine and one-half hours. This was in a factory where all the tires were welded and the wheels ready so that it was nothing but to heat the tires and put them on. I had three fires with twelve tires in each fire. An artesian well running through the water box kept the water cool.

If the fire is not hot enough to make it expand a tire puller is needed. A tire puller can be made in many ways and of either wood or iron. Buggy tire is more particular than wagon tire and there are thousand of buggy wheels spoiled every year by poor or careless blacksmiths. In a buggy tire one-eighth of an inch draw is the most that it will stand, while most wheels will stand only one-sixteenth. If the wheel is badly dished don't give it any draw at all, the tire should then measure the same as the wheel, the heat in the tire is enough.

If the wheel is fellow-bound cut the fellows to let them down on the spokes.

If the spokes are loose on the tenon wedge them up tight.

BACK DISHED WHEEL

For a back dished wheel a screw should be used to set the wheel right. Place the wheel on the table front side up. Put wood blocks under the fellow to raise the wheel up from the table. Place a two by four over the hole under the table; have a bolt long enough to reach through the two by four and up through the hub, a piece of wood over the hub for the bolt to go through; screw it down with a tail nut. When the wheel is right, put the tire on. The tire for such a wheel should have more draw than for a wheel that is right.

If a buggy wheel has been dished it can be helped a little without taking the tire off. Place the wheel on the anvil so that the tire will rest against the anvil. Don't let the tire rest lengthwise on the anvil. If you do, the tire will be bent out of shape when you begin to hammer on it. Use the least surface possible of the anvil and hammer on the edge of the tire; the stroke of the hammer to be such that the blow will draw the tire out from the fellow. A tire too tight can be remedied this way.

When bolting a wheel the tire will be out of place unless the tire has been shrunk alike on both sides of the fellow plates. A smith used to setting tires will be able to get the holes almost to a perfect fit. If a tire is too short, don't stretch it with a sharp fuller that will cut down into the tire, when the tire is a little worn it will break in this cut. Draw it out with a wide fuller and smooth it down with the hammer. If it is much too short, weld in a piece. This is easily done. Take a piece of iron ¼-inch thick, the width of the tire and the length needed, say about three inches. Taper the ends and heat it to a red heat. Place it on the tire in the fire and weld. This will give material for stretching.

If the wheel has a strong back dish it cannot be set right to stay with the tire alone, as a bump against the fellow is apt to throw the dish back. It is therefore safer in all back dished wheels to take the spokes out of the hole and set them right by wedges in the end of the spokes. These wedges should not be driven from outside in but be placed in the end of the spoke so that they will wedge into the spoke when the same is driven back into its place. Use glue.

HOW TO PUT ON NEW TIRE

When you have the bar of either steel or iron for the tire, first see if it is straight, if not be sure to make it. Next place the tire on the floor and place the wheel on top of the tire, begin in such a way that the end of the fellow will be even with the end of the tire. Now roll the wheel over the tire. If a heavy tire cut it three inches longer than the wheel, if a thin tire, two inches. Now bend the tire in the bender. Measure the wheel with the gauge, then measure the tire; if it is a heavy wagon tire and a straight wheel cut the tire one-fourth of an inch shorter than the wheel. If it is a buggy tire cut it the size of the wheel. In welding these tires they will shorten enough to be the size wanted.

HOW TO WELD TIRES

There are many different ideas practiced in welding tires. One smith will narrow both ends before welding; another will cut the edges off after it is welded. This is done to prevent it from spreading or getting too wide over the weld. I hold that both these ideas are wrong. The first one is wrong because when the ends are narrowed down it is impossible to make them stay together until the weld is taken, especially if it is a narrow tire. The second idea is wrong because it cuts off the best part of the weld and weakens it. Some smiths will split the tires, others will rivet them together. This is done to hold the tire in place until it has been welded. There is no need of this trouble, but for a new beginner a rivet is all right.

I shall now give my experience in welding tire, and as this experience has been in a factory where thousands of wheels are made yearly, I suppose it will be worth something to the reader.

When the tire is ready to weld draw down the ends and let them swell as much as they want to. Now let the helper take the end that is to lay on top and pull it towards the floor, the other end to rest on the anvil. This will give that end a tendency to press itself steadily against the lower end. Next place this end on top of the other end. The ends must now be hot enough to allow them to be shaped. You will now notice that the top end is wider than the tire, so is the lower end. The tire is to be so placed that the swelled parts reach over and inside of each other a little. **Now** give a couple of blows right over the end of the under tire. Next tap the swelled sides down over the tire. This will hold the tire together so that it cannot slip to either side, and the swelled end of the under tire will prevent it from pulling out. If the top end has been so bent that it has a tendency to press down and out a little, the tire will now be in a good shape to weld.

Before you put the tire into the fire, let me remind you of what I have said before about the fire. Many blacksmiths are never able to weld a tire tight on the outside because of a poor, low, and unclean fire. If the fire is too old or too fresh it will not give a good heat for welding tire. If you have a good big fire high up from the tuyer, then you are all right. Place your tire in the fire and proceed as follows: No matter whether it is an iron or soft steel tire, sand is the best welding compound and nothing else should be used; but if you lose the first heat then borax might be used as it will prevent the tire from scaling and burning. When you have the right heat, place the tire on the anvil this way; let the tire rest against the inside edge of the anvil. If the lower end of the tire is allowed to come down on the anvil it will cool off and can never be welded that way. Now hold the tire this way until you have the hammer ready to give the first blow. Then let the tire down and strike the first blows directly on top and over the end of the under end. This is important and if the first blows are not directed to this very place the lower end will be too cool to weld when you get to it. Next weld down the upper end, this done turn the tire on edge and while it is in a welding heat come down on it heavy with hammer, if a buggy tire, and with a sledge and hammer if a heavy wagon tire. Hammer it down until it is considerably narrower over the weld as it will swell out when dressed down. This way the weld has all the material in the iron and the lapped lips will help hold the weld together. A very poor smith can weld tires to stay in this manner. The edges should be rounded off with the hammer and filed to make the tire look the same over the weld as in the iron. If there should be any trouble to weld a steel tire place a little steel borings over the weld and use borax.

A blacksmith in Silver Lake, Minn., working for a wagon maker of that place, when welding a tire failed entirely after half a dozen attempts, and he got so angry that he threw the tire down on the floor with all his might. It happened to crush the wagon makers big toe. This was more than the otherwise good-natured man could stand, and instantly the smith was seen hurled through an open window—the wagon maker attached. Result: separation and law suit. All this because the smith had not read my book.

When a light buggy tire is to be set mistakes are often made in measuring the tire. The tire is too light in itself to resist the pressure of the gauge. The smith tries to go it light and if there is not the same pressure in measuring the tire there was in measuring the wheel, it will not give the same results; and when the tire is put on it is either too tight or too loose. I worked for many years on a tool to hold the tire steady in order to overcome this trouble. The only device that I have ever seen for this purpose before is the anvil close up to the forge, one side of the tire on the forge, the other on the anvil. This arrangement would crowd the smith, roast his back and expose him to ridicule, but it will not help to ruin the tire.

The tool I invented is a tire holder made of cast iron. It consists of a standard or frame with a shank in to fit in the square hole in the anvil; in the standard is a slot hole from the bottom up. On the back of the standard are cogs on both sides of the slot hole. Through this goes a clutch hub with cogs in to correspond with the cogs in the standard. On the outside of the standard is an eccentric lever. Through this lever is a tapered hole to fit over the clutch hub. This lever is tapered so that it will fit different thicknesses, while the cogs and eccentric lever will adjust it to different widths. This device is so cheap that any smith can afford to have it.

Next time you buy a quart of whisky sit down and figure out which will do you more good, my tire holder or the whisky. Figure 7 is an illustration of my holder. This tool is better than an advertisement in your local paper, of which the following story will convince you. A blacksmith in a prohibition county in a northern Iowa town got into the habit of going over to a Minnesota town for a keg of beer every month. On one of his periodical visits to this place he saw a crowd of men standing around a road grader in the road. As he approached he found that the grader had a serious break-down and the men were

MODERN BLACKSMITHING 81

just discussing the possibilities of getting the grader repaired in the village shops. One said no smith could do it, another thought they could if they only had tools. "I know a man," said one in the party,

Fig. 7

HOLMSTROM TIRE HOLDER

"that can if any man can, and he has tools I am sure. I was over to his shop the other day to have my buggy tire set, and mind you, he had the slickest tool you ever saw to hold the tire in; I never saw a tool like that before." "Well," said one, "that has nothing to do with this case." "Yes it has," said the road boss, "my father always used to say, 'A mechanic is known by the tools he uses,' and when a smith has good tools in one line, he has them in another, and I shall give this man a chance."

Our traveling smith had heard enough. This was a temperance and tool lecture to him, he began to think of all the trips he had made to this town. Twelve trips a year, three dollars a trip for liquor and the time lost must be worth two dollars per day. He figured it out and would have turned back if he had not been so close to the place. He took a glass of beer but it didn't taste as usual and he asked for a cigar. With this he returned, and on the road home swore off for good. He bought a tire holder at once to start in with, and by this time he is one of the best smiths in the country, always at his stand ready to do the work brought to him, and his customers now know that he is to be found in his place, with tools of all kinds and a sober hand to use them with. Do thou likewise.

TIRE IN SECTIONS

Many of us remember the time when tires were made in sections and nailed on, at this time the wheels were more substantially made, because the tire could not be set as tight as it is now, and the wheel had to be made so that it would stand the usage almost independent of the tire. Our endless tire is a great improvement over the tires made in sections. The wagon tires as they are made now are, I think, as near right as they can be, in regard to size of iron, in proportion to the wheel. But it is different with buggy tires. I hold that they are all made too light to be of any protection to the fellows. I understand the reason why they are made this way, but if a man wants a light rig, let that be the exception and not the rule.

Tire should not be less than one-fourth of an inch thick for seven-eighths wide, and five-sixteenths for an inch wide and over.

EXPANSION OF THE TIRE

A tire four feet in diameter will expand two inches and a quarter, or three-sixteenths of an inch to the foot. Steel tire expands less. This is the expansion of red heat. If heated less it expands less, but it is no trouble to make the tire expand for all the draw it needs.

A furnace for tire heating comes handy in cities where there is no chance for making a fire outside, but every smith that has room for a fire outside will do better to heat the tire that way. Don't build a tire-heating furnace in the shop if wood is to be used for fuel, because the heat and smoke will turn in your face as soon as the doors of the furnace are opened.

WELDING AXLES

When a worn buggy axle is to be stubbed, proceed as follows: First, measure the length of the old axle. For this purpose take a quarter inch rod of iron, bend a square bend about an inch long on one end. With this rod measure from the end of the bearing, that is, let the hook of your rod catch against the shoulder at the end where the thread begins, not against the collars, for they are worn, nor should you measure from the end of the axle, for the threaded part is not of the same length. Now place your stub on the end of the axle and mark it where you want to cut it off. Cut the axle one-fourth inch longer than it should be when finished. Next heat the ends to be welded and upset them so that they are considerably thicker over the weld; lap the ends like No. 1, Figure 6, weld and use sand, but if the ends should not be welded very well then use borax. These stubs are made of soft steel, and will stand a higher heat than tool steel, but remember it is steel. If the ends have been upset enough they will have stock enough to draw down on, and be of the right length. If this is rightly done one cannot tell where the weld is. Set the axle by the gauge, if you have one, if not, by the wheels.

AXLE GAUGE

A gauge to set axles by can be made in this way: When you have set an axle by the wheels so that it is right, take a piece of iron 1¼ x ¼, six feet long, bend a foot on this about six inches long, with a leg on the other end. See No. 5, Figure 8; the leg to be movable and set either with a wedge or a set screw to fit for wide and narrow track. The gauge to be set against the bottom side of the axle. The pitch to be given a set of buggy wheels should be from one to one and one-half inches. I would recommend one and a half inches. This will be enough to insure a plumb spoke when the vehicle is loaded. It will also insure safety to the rider from mud slinging. By pitch, I mean that the wheels are one and a half inches wider at the upper rim than they are down at the ground. Every smith ought to have a gauge of this kind, it is easy to make and it saves a lot of work, as there is no use of the wheels being put on and an endless measuring in order to get the axle set right.

GATHER GAUGE

By gather I mean that the wheels should be from one-fourth to one-half an inch wider back than in front. Don't misunderstand me now. I don't mean that the hind wheels should be wider than the front wheels, I mean that a wheel should have a little gather in front, as they are inclined to spread and throw the

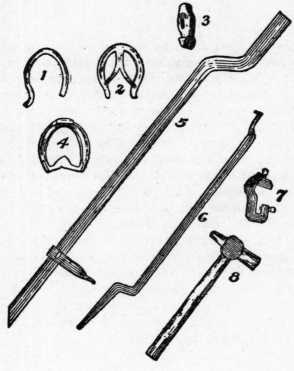

Fig. 8

bearing on the nut, while, if they have a little gather, they will run right, and have a tendency to throw the bearing on the collars of the axle. If they do they will run more steady, especially when the axle is a little worn.

A gauge for this purpose can be made like Figure 8, No. 6. This gauge to be fitted to the front side of the axle when you make it. It can be made of 1 x ¼ about three feet long, the forked end to reach the center of the axle. With these two gauges axles can be set right without the wheels.

CHAPTER V

HOW TO MAKE PLOWSHARES

HERE are two kinds of shares: lip shares and bar shares, and they must be treated differently. We will first treat of bar shares. The first thing to do when a plow is brought for a new lay is to look over the condition of the landside. By landside is meant the bar to which the share is welded. Now if this bar is worn down so that you think it too weak to stand for a new share, then make a new one.

HOW TO MAKE A LANDSIDE

For a 14-inch plow take $2\frac{1}{2} \times \frac{3}{8}$, or $2\frac{1}{2} \times \frac{7}{16}$. For a 16-inch plow, use $2\frac{1}{2} \times \frac{7}{16}$, or $3 \times \frac{7}{16}$ common iron. Cut the iron diagonally at the point. This will prepare a point on each side of the cut; that is, you had better cut out two landsides at a time. But if you do not want to do that, then cut the iron off square. Next take a piece of common iron $3 \times \frac{1}{4}$, 13 inches long for a shin; cut this diagonally, and it will make shins for two. Some plow factories use steel for shins, but that is not necessary, for it will not make the plowshare any better, but, on the other hand, will be quite a

JOHN DEERE, THE INVENTOR OF STEEL PLOWS

bother when you want to drill a hole for a fincoulter if it is hardened. Place this shin on the land side of the landside, and weld. In preparing the shoulder of the shin for the plate use a ship upsetter. See No. 3, Figure 8.

Not one out of 500 blacksmiths have this tool. Every smith should have one. You cannot do a good and quick job without it.

When you shape the point of the landside hold it vertical, that is, the edge straight up and down, or plumb. If you don't do this, there will be trouble in welding, especially if you have held it under. Then it will lean under the square when welded, and in such a case it is hard to get a good weld, and if you do you will break it up when you attempt to set it to the square. Another thing, don't make much slant on the landside up at the joint, for, if you do, you can never weld the share good up there. Give more slant towards the point. Be sure to have the right curve. It is very important to have the landside right: 1, Because it is the foundation for the plow; 2, if the landside is right the start is right, and then there is no trouble to get the share right. When finished place the old landside on top of the new, with the upper edges even; don't go by the bottom edges, as they are worn. Now mark the hole. You may leave the front hole for the foot of the beam this time. When holes are drilled, then put a bolt through the hole of the foot of beam and landside; now place the plow on the landside and measure 14 inches from the floor up to the beam. In this position mark the front hole of the foot of the beam. If the beam has been sprung up you will now have remedied that. So much about **a new land**side. On the other side, if the old landside **is not** too much worn to be used, then repair as follows: Take a piece of ⅜-inch thick flat iron the width of the landside about ten inches long. Cut one end off diagonally, this end to be flattened down. Why should this end be cut diagonally? This piece of iron is to be placed on the inner side of the landside and as far back as to cover the hole that holds the plate. Now, if this iron is cut square off, and left a little too thick on that end, it will cut into the landside and weaken it; but if cut diagonally and drawn out thin it will not weaken, nor can it break when cut in this manner. To be sure of a good strong weld, upset over the weld. I hold that this is the most important thing in making a new lay. "No hoof, no horse"— no landside, no plow. There are only a few blacksmiths recognizing this fact. Most of the smiths will simply take a piece of iron about half an inch square and weld it on top of the point. This is the quickest way, but it is also the poorest way, but they cannot very well do it in any other way, for if you have no shin upsetter to dress and shape the shoulder for the plate, then it is quite a job to repair any other way. There are three reasons why a landside cannot be repaired with a patch on top of the point: 1, The shin or shoulder in an old landside is worn down sometimes to almost nothing, and the only way to get stock enough to make a good shoulder is to put a good-sized piece of iron on the inside, back and behind this shoulder. If a new plate is to be put on and this is not done, you will have to draw down the plate to the thickness of the old shoulder, and in such a case the plate will add no strength to the share.

2, The landside is, in many cases, worn down on the bottom to a thin, sharp edge, and by placing the piece on top the landside will be as it was on the bottom side, where it ought to be as thick as you can make it.
3, The weakest place in the landside is just at the shoulder of the shin, and by placing the piece on top it will not reach over this weak place, and with a new long point on, the strain will be heavier than before, and the landside will either bend or break. I have in my experience had thousands of plows that have been broken or bent on account of a poorly-repaired landside. Blacksmiths, with only a few exceptions, are all making this mistake.

The landside is to the plow what the foundation is to the house. No architect will ever think of building a substantial house without a solid foundation. No practical plowsmith will ever try to make a good plow without a solid landside.

For prairie or brush breakers, where no plate is used, it will be all right to repair the landside by placing a piece of iron on top of it, provided it is not much worn, and the patch reaches back far enough to strengthen the landside. But even in such cases it is better to lay it on the inner side.

LANDSIDE POINT FOR SLIPSHARE

We have now learned how to prepare the landside for a solid or long bar share. We shall now learn how to make a landside point for slipshares. There are smiths that will take the old worn-out stub of a slip-share point, weld a piece to it, and then weld the share on. This is very ridiculous and silly. There is nothing left in such a point to be of any use. Make a new one; be sure to make it high enough—at least half an inch higher than the share is to be when finished. This will give you material to weld down on. If the landside is not high enough the share will be lower—that is, the joint of the lay will be lower than the joint of the mouldboard, and it should be the other way.

PLOW OF 200 YEARS AGO

On this point many an old smith and every beginner makes mistakes, and not only in this case, but in everything else. Whatever you have to make, be sure to have stock enough to work down on, and you will be all right. It is better to have too much than not enough.

In shaping the point remember to hold it perpendicular, and give very little slant up at the joint, but more towards the point. If too much slant up at the joint there will be difficulty in welding it. Remember this. Don't make the point straight like a wedge; if you do the share will be above the frog. Give it the same circle it had, and the share will rest solid on the frog. This is another important point to remember:

The lay will not have the full strength if it don't rest on the frog, and it will not be steady, and the plow will not run good, for in a few days the share flops up and down.

When a 14-inch share is finished the point, from the joint of the share to the extreme end of the point, should be 11 inches, not longer, and for a 16-inch lay, 12 inches, not longer. The point acts as a lever on the plow, and if it is too long the plow will not work good, and it is liable to break. Shape the point so that when you hold it up against the plow it will be in line with the bottom of the landside, but about half an inch wider than the landside to weld on. If it is a plow where the point of the mouldboard rests on the landside point, and it is a double shin, then cut out in the landside point for the point of the mouldboard to rest in. See No. 1, Figure 9. This will be a guide for you when welding the share, and it will slip onto the plow easier when you come to fit it to the same. I think enough has been said about the landside to give the beginner a good idea of how to make one. And if the landside is right, it comes easier to do the rest. In making a plowshare there are many things to remember, and one must be on the alert right along, for it will give lots of trouble if any point is overlooked.

We will now weld a share to a long bar landside. The landside having been finished and bolted to the beam or its foot, or to a standard, the share is to be shaped to fit. Hold the share up to the plow. First look if the angle for the point is right in the share; if not, heat the share, and if under the angle wanted upset up at the joint; if over the angle wanted, drive it back at the point. In doing this hold the edge of the share over a wooden block instead of the anvil, so as not to batter the thin edge of the share. If the share has been upset so that it has a narrow rib along the point where it is to be welded, draw this down and make it level. In most blank shares the point should

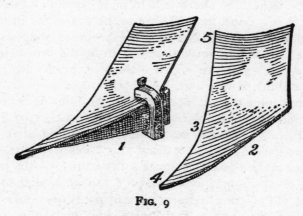

FIG. 9

be raised to fit the landside point, so that when the same is placed on the floor the edge of the share will follow the floor or leveling block (if you have it), from the heel right up to the point, then it will be easy to make the edge come down to the square in finishing it up. If this is not done the edge of the share from the throat back will generally be too high.

In Figure 9 two shares are represented, one with the landside point on ready for welding. In this share the point of the same has been raised so that the share comes down to the square in the throat. The other is a blank share, straight in the point between Nos. 4 and 5, resting on the extreme heel and point with gap between the edge of share and floor at No. 3. In most blank shares the point is too straight, and the point too much bent down at No. 4. Bend the share so that the whole length from heel to point will follow the floor. When the share is held in a position as shown in this cut, don't fit the share to the brace, for in most old plows the brace has been bent out of shape. Fit the share to the square, and then fit the brace to the share, and you are right. Many a blacksmith will never think of this, but it is important.

Next joint the share; that is, if the joint does not fit the joint of the mouldboard, make it fit either by filing or grinding. This done, make the holes, and when you center-punch for same draw the holes a trifle; that means make the center mark a little towards the inner side of the mark, especially for the hole next to the point. This is also an important point overlooked by most blacksmiths. The holes that hold the joints together should act as a wedge. If they don't the joints will pull apart and leave a gap between, where dirt and straw will gather, and if a slipshare the share will soon work loose and the plow will flop.

The holes having been punched and countersunk, the share should be bolted to the brace. Next put on the clamp. It is not necessary that the clamp should be put on while the share is on the plow. I never do that. I used to for many years, but there is no need of doing it, for if the share has the right angle it must come to its place when even with the point on the outside, and a cut should be made in the landside just at the place where the point of the mouldboard rests on same, this cut will also be a guide.

Now a few words concerning the clamp. Figure 8, No. 7 illustrates a clamp for this purpose. The set screw at the bottom serves to hold the landside from leaning over or under, while the setscrew at the upper end holds the share against the point. If this clamp is rightly made it works splendid. The clamp should be placed over the plowshare up at the joint, because the first heat or weld should be on the point. Some smiths —well, for a fact, most smiths—take the first weld up at the joint. This is wrong. The point should be welded first. Then you have a chance to set the share right and fit it snug to the point the whole way up. You cannot make a good weld if the share does not fit snug against the landside point, to prevent air and cinders from playing between. Further, the share should be upset over the weld, when this is not done in the blank share; the lower corner of the share will protrude over the landside. This should be dressed down smooth. The next weld should be taken up at the joint. For welding compound use steel borings and scales from either steel or iron.

After you have moistened the place where the weld is to be taken with borax, then fill in between the share and point with steel borings, and on top of this a little steel or iron scales. Do not buy any welding compound of any kind, because if you learn to know what you have in the shop you will find that there never was a welding compound made to excel borax, steel scales, steel or iron borings, and powdered glass. All these you have without buying.

In heating go slow. If you put on too strong blast the share will burn before the iron is hot enough to weld. When ready to weld let your helper take with a pair of tongs over the share and landside to hold them tight together while you strike the first blow. Use a large hammer and strike with a pressure on the hammer the first blows, until you are sure it sticks; then come down on it with force.

I have made it a practice, no matter how good this weld seems to be, to always take a second weld. This weld to be a light one. The share and landside are after the first weld settled, so it takes very little to weld them then. On the other hand, the first weld might look to all appearances solid, but it is not always. With this precaution I never had a share that ripped open in the weld, while it is a rare thing to find a share made by a blacksmith that does not rip. Now, then, weld down toward the point. The point should not be allowed to have any twist, for if it does, it will turn the plow over on the side. Now set the edge right, beginning at the heel. If the share is made for hard fall plowing give more suction than for a share for soft spring plowing. Grind and polish before you harden, and after it is hardened touch it up lightly with the polish wheel. Much polishing or grinding after hardening will wear off the case hardening.

SLIPSHARE

We shall now weld a slipshare. When the point is finished hold it to the plow with a pair of tongs while you fit the share. When the share is fitted take the point off from the plow and fasten it to the share with the clamp. As I have said before, there is no need of fastening the share to the landside point with the plow as a guide. If the landside and share are right there cannot be any mistake, and it comes easier to screw them together over the anvil. Now proceed as with a long bar share, and when the weld up at the joint has been taken, fit the share to the plow while hot. Some smiths in preparing the landside point for a slipshare will place the share so that the point is a little too short back where it rests against the end of the plate. This is a bad idea. It is claimed that, in welding, the landside point will swell enough to make it reach up against the plate. This is true, if the landside point is only high enough; but if it is low and you lose a heat in welding, as most smiths do, then your landside point will be both too low and too short. Thousands of shares are made every year that have this fault. Therefore, whatever you are doing have stock enough. It is easy to cut off from the landside while yet hot, but it is difficult to repair if too short. No share will work steadily if the point does not rest right against the plate.

In blacksmithing, every beginner, and many an old smith, makes the mistake of providing less stock than is needed for the work to be done. It is essential to have material to dress down on; and if a heat is lost, or a weld, it will make the stock in the article weaker, and to meet these exigencies there must be material from the start, enough for all purposes. There is also

a wide difference of opinion as to whether the share should be welded at the point or at the joint first. While I was yet a young man and employed in a plow factory, I had an opportunity to see the different ideas set to a test. In the factory the practice was to weld the point first. A plowman from another State was engaged, and he claimed that it would be better to weld the share first up at the joint. He was given a chance to prove his assertion, and the result was that 3 per cent of his shares broke over the inner side of the landside at the joint in the hardening, and 10 per cent ripped up in the weld at the same place. These are results that will always follow this method.

The first, because the share was not upset over the weld; the second, because a good weld cannot be taken unless the share is dressed down snug against the point when hot. As far as the number of shares welded per day was concerned, this man was not in it. Still, this man was a good plowman, and was doing better than I ever saw a man with this idea do before. For it is a fact, that out of one thousand plowshares welded by country blacksmiths, nine hundred and ninety will rip up. I have been in different States, and seen more than many have of this kind of work, but, to tell the truth, there is no profession or trade where there is so much poor work done as in blacksmithing, and especially in plow work. Blacksmiths often come to me, even from other States, to learn my ideas of making plowshares. On inquiring, I generally find that they weld a piece on the top of the old landside and proceed to weld without touching the share or trying to fit it at all. We need not be surprised at this ignorance, when we know that it is only fifty years since John Deere reformed the plow industry entirely and made the modern plow now in use. It is impossible for blacksmiths in the country to have learned this part of their business, in so short a time, successfully. Still, I have seen blacksmiths prosper and have quite a reputation as plowmen, while, for a fact, they never made a plowshare that was, from the standpoint of a practical plowman, right.

CHAPTER VI

HOW TO HARDEN A PLOWSHARE

the share is of soft center steel, harden as follows: First, heat the whole point to a very low red heat; then turn the share face down, with the heel over the fire, and the point in such a position that it is about two inches higher than the heel. This will draw the fire from the heel along towards the point, and the whole length of the share will be heated almost in one heat. Be sure to get an even heat, for it will warp or crack if the heat is uneven. When the share has a moderate red heat take it out, and you will notice that it is sprung up along the edge. This is the rule, but there are exceptions, and the share is then sprung down. In either case set it right; if sprung up set it down a little under the square; if sprung down set it a little over the square. You cannot with any success set it by a table or leveling block, because this will, first, cool off the edge, second, it must be either over or under the square a little. Therefore, you must use your eye and set the share with the hammer over the anvil. This done, hold the share over the fire until it has a low red heat, as stated before; then plunge it into a tub of hardening compound, such as is sold by the traveling man, or sprinkle the share with prussiate of potash and plunge it into a barrel of salt water.

You will notice that the share will warp or spring out of shape more in the heating than it does in the cooling, if the heat is right. Some smiths never look at the share when hot for hardening, but simply plunge it into the tub, and then they say it warped in hardening, while it was in the heating. If the share is too hot it will warp in cooling also.

HOW TO POINT A SHARE

Points are now sold by dealers in hardware, and every smith knows how they are shaped. There is, however, no need of buying these; every smith has old plowshares from which points can be cut, provided you don't use an old share too much worn. The points sold are cut with the intention that most of the point is to be placed on top of the plow point. This is all right in some instances, while it is wrong in others. When you cut a piece for a point make it the same shape at both ends. Now, when a plow needs the most of the point on top bend the end to be on top longer than the end to go underneath, and vice versa, when the point wants to be heaviest on the bottom side. I hold that in ordinary cases the most of the point should be on the bottom side. If it is it will wear better and keep in the ground longer, for as soon as the point is worn off underneath it comes out of the ground.

Don't monkey with old mower sections or anything like them for points, for, although the material is good, it is not the quality alone but also the quantity that

JAPANESE PLOW

goes to make up a good point. It takes only a few hours' plowing to wear off a section from the extreme point of the share, and then there is only the iron of the plow point left to wear against, and your time spent for such a point is lost. Another thing, it takes just as much time to put on such a point as it does to put on a good one for which you charge the regular price.

In putting on a point of thin material you must go unusually slow, or you will burn the steel before the plow point is hot.

Smiths, as a rule, draw out a round back point. They seem to be afraid of coming down on the point with the hammer for fear it will spring the point towards the land. This can be remedied by using a wooden block for anvil. Then you can set the point back without battering the edge of the share. The

BENCH FOR HOLDING PLOWS WITHOUT BEAM

suck of a point should be one-eighth of an inch. Don't split the steel of the point of a share open and wedge a point in. Make one long enough to reach around the point, say from 8 to 10 inches long, and you will have a good substantial job. There is too much experimenting in putting on points yet, but the method just described is the only good one.

HOW TO SHARPEN A PLOWSHARE

If the share to be sharpened is a hardened share, and it is the first time it is sharpened, then be careful not to heat it too far towards the joint, so as to leave the temper as much the same as possible. For my part, I never follow this rule. I heat it as much as is needed to draw it out good, and then harden it over again. But beginners can sharpen a new share once without hardening it over, if the temper is not entirely out of the share. To sharpen a share without springing it some is an impossibility. No device will prevent this, and the only way to set it right is to heat it all over. In sharpening a share it is drawn out on one side, and it is natural that that side is made longer, and as a result the share must warp. In a circular saw it takes only a couple of blows on one side to get it out of shape; then what else can we expect in a plowshare, when all the hammering is done on one side?

Some smiths turn the bottom side of the share up and hammer on that side, but this is wrong; first, because in so doing you unshape the share; second, the scales on the anvil will mark the face of the share just as bad as the hammer, so nothing is gained by this. Place the share on the anvil, face up, and use a hammer with a big round face, and when you get used to this, the best result is obtained. Don't draw the edge out too thin. There is no need of a thin edge on a plow that has to cut gravel and snags, but for sod breaking a thin edge is wanted, and the smith has to use his best judgment even in such a case.

HOW TO PUT ON A HEEL

Cut a piece of steel about eight inches long, three inches wide on one end, and pointed down to a sharp point on the other. Draw out one side thin to nothing. Next, draw out the heel of the share. Now place the heel piece on the bottom side of the share, and hold it in place with a pair of tongs and tong rings. Take the first heat at the pointed end of the piece, next heat at the heel, share down, then turn the share over, heel down; go slow, use borax freely, and place a little steel borings between the heel piece and the share. After a little practice almost any smith ought to be able to put on a heel, while now it is only a few smiths that can do it. I never put on a heel yet but the owner of the plow would tell me that other smiths tell him it cannot be done. When welded good be sure to get the right shape in the share. Grind and polish carefully, as the dirt is inclined to stick to the share in this place more easily than in any other.

HOW TO REPAIR A FLOPPING PLOW

When a plow is flopping or going everywhere so that the owner don't know what is the matter the fault should be looked for first in the beam. If the beam is loose the plow will not run steady, but the reason for this trouble, in most cases, is in the share. If the point has too little "suction," and the edge of the share is too much rolling the plow generally acts this way.

To remedy this, sharpen the share, set the point down, and the edge of the lay from the point all the way back to the heel, and the plow will work right.

HOW TO SET A PLOW RIGHT THAT TIPS ON ONE SIDE

If a plow is inclined to fall over on the right handle, the fault is in the share. The share in such a case has too much suction along the edge. Heat the whole share and roll the edge of it up and the plow will work all right.

If a plow tips over on the left side handle, the share in such a case is too much rolled up. Heat it all over and set the edge down to give it more suction.

WHEN A PLOW RUNS TOO DEEP

There are two reasons for a plow running too deep: 1, If the beam is more than fourteen inches high from the floor up to the lower side of it, then the beam should be heated over a place as far back as possible, and the same set down to its proper place. 2, If the point of the share has too much suction the plow will also run too deep. The right suction to give a plowshare is from $\frac{1}{8}$ to $\frac{3}{16}$ of an inch. If a plow don't run deep enough with this much as a draw, there must be something else out of shape; or, if it goes too deep, the fault must be looked for in the beam or in the tugs with small-sized horses. The point of a share should never be bent upwards in order to prevent the plow from going too deep. Set the share right, and if the plow then goes out of its proper way the fault must be found somewhere else.

WHEN A PLOW TAKES TOO MUCH LAND

If a 14-inch plow takes too much land the fault is either in the point of the share or in the beam. The point of a share should stand one-eighth of an inch to land, and the beam should stand about three inches to the right. This will be right for a 14-inch plow and two horses. If for a 16-inch plow and three horses, the beam should be in line with the landside.

HOW TO FIX A GANG PLOW THAT RUNS ON ITS NOSE

When a gang or sulky plow runs on its nose and shoves itself through the dirt, the fault is with the share or in the beam. In most cases this fault is a set back beam, but it might also be the result of a badly-bent-down and out-of-shape landside point. If it is in the beam, take it out and heat it in the arch, then bend it forward until the plow has the right shape, and it will run right.

HOW TO HARDEN A MOULDBOARD

To harden a mouldboard is no easy job in a blacksmith's forge, and it is no use trying this in a portable forge, because there is not room enough for the fire required for this purpose. First, dig the firepot out clean, then make a charcoal fire of two bushels of this coal, have some dry basswood or wood like it, and when the charcoal begins to get red all over then pile the wood on the outside corners of the fire. Heat the point of the mouldboard first, because this being shinned, it is thicker and must be heated first or it will not be hot enough; then hold the mouldboard on the fire and pile the wood and hot coal on top of it. Keep it only until red hot in the same place, then move it around, especially so that the edges get the force of the fire, or they will be yet cold while the center might be too hot.

HOW TO PATCH A MOULDBOARD

When the mouldboard is red hot all over sprinkle with prussiate of potash, and plunge into a barrel of ice or salt water. A mouldboard will stand a good heat if the heat is even; otherwise it will warp or crack. Another way to heat a mouldboard: if you have a boiler, then fill the fire place with wood and heat your mouldboard there. This will give you a very good heat. If it is a shinned mouldboard the point must be heated first in the forge, then place it under the boiler for heating. This must be done to insure a good heat on the point, which is thicker than the mouldboard and therefore would not be hot enough in the time the other parts get hot.

When a mouldboard is worn out on the point a patch can be put on, if the mouldboard is not too much worn otherwise. Cut a piece of soft center steel to fit over the part to be repaired. Draw this piece out thin where it is to be welded to face of mouldboard. Hold

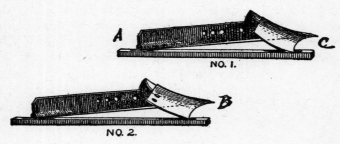

FIG. 11

this piece in position while taking the first weld, with a pair of tongs. Weld the point first, then the edges, last the center. The patch should be welded to face of mouldboard. When the last weld is taken place the mouldboard face up, with some live coal over it, in the fire; use borax freely, and, when ready to weld, weld the patch while the mouldboard is in the fire, using a ⅜ rod of round iron as a hammer with one end of it bent for this purpose. When the patch is thus welded in its thinnest place then take it out and weld on the anvil. In heating for the weld never place the patch down towards the tuyer, for there the blast will make it scale, and it will never weld this way. Remember this in all kinds of welding.

Figure 11A represents two shares. No. 1 represents a share set for spring plowing, when the ground is soft. Notice the heel of the share following the square for about one inch at c, while the heel in No. 2 rests with the extreme edge on the square, and is set for fall plowing, when the ground is hard. The line between a and b shows the suction at d, which is not more than an eighth of an inch. Breaking plows and large plows which are run shallow should have a wide bearing at c. In breaking plows the heel will sometimes have to be rolled up a little at this place.

CHAPTER VII

MOWER SECTIONS

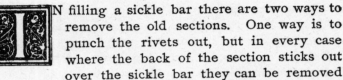IN filling a sickle bar there are two ways to remove the old sections. One way is to punch the rivets out, but in every case where the back of the section sticks out over the sickle bar they can be removed easier in this way: Just open the vise enough to receive the section, then strike with the hammer on the back of the section, and this blow will cut the rivets off. You can cut out ten to one by this method to any other.

Sometimes the sickle bar is bent out of shape in the fitting. To straighten it place the sickle on the anvil, sections down; now strike with the hammer so that it will touch the bar only on one half of its face, the blow to be on the inner side of the curve.

BABBITING

When a box is to be babbited the first thing to do is to clean the box If it can be placed over the fire the old babbit with melt out easily. If the box cannot be held over the fire, then chisel the old babbit out. At each end of the box there is a ridge to hold the babbit in the box; that is, in cast iron boxes. On top of this ridge place a strip of leather as thick as you want the babbit to be. This done, place the shaft in the box. Pour the babbit in level with the box. Be careful about having the box dry; if any dampness is in the box the babbit will explode. Now place a thick paper on each side of the box and put on the top box, with the bolts in to hold it in place tight, then close up at the ends with putty. In some cases it is best to heat the box a little, for if the box is cold and there is little room for the babbit it will cool off before it can float around. In such a case the boxes should be warm and the babbit heated to a red heat. Now pour the babbit in through the oil hole.

In cases where there are wooden boxes, and the babbit is to reach out against the collars, the shaft must be elevated or hung on pieces of boards on each side with notches in for the shaft to rest in. Use putty to fill up and make tight, so that the babbit must stay where wanted. For slow motion babbit with a less-cooling percentage (tin); for high speed, more-cooling (tin). Grooves may be cut in the bottom box for oil. When a shaft is to be babbited all around in a solid box the shaft is inclined to stick in the babbit. To prevent this smoke the shaft a little and have it warm. When cool it will come out all right. Or wind thin paper around the shaft, the paper to be tied with strings to the shaft.

ANNEALING

By gradually heating and cooling steel will be softened, brittleness reduced, and flexibility increased. In this state steel is tough and easiest drilled or filed. Tool steel is sometimes too hard to drill or file without first annealing it; and the best way to do this is to slowly heat to a red heat, then bury the steel in the cinders and let it cool slowly. To heat and let the steel cool exposed to the air will do no good, as it cools off too quick, and when cool the steel is as hard as ever. This is air temper.

HOW TO REPAIR BROKEN COGS

Cogs can be inserted in a cogwheel in different ways. If the rim of the wheel is thick enough a cog can be dovetailed in. That is, cut a slot in the rim from the root of the cog down, this slot to be wider at the bottom. Prepare a cog the exact size of the cogs, but just as much deeper as the slot. Before you drive this cog in, cut out a chip on each end of the slot, and when the cog is driven in you can clinch the ends where you cut out. This will make a strong cog, and if properly made will never get loose.

Another way: If the rim is thin, then make a cog with a shank on, or a bolt cog. If the rim is wide make two bolts. The cog can be either riveted or fastened with nuts. If only one shank is made, the same must be square up at the cog, or the cog will turn and cause a breakdown. But a shallow slot can be cut in the rim to receive and hold the cog, and then a bolt shank will hold it in place, whether the shank is round or square.

HOW TO RESTORE OVERHEATED STEEL

If steel has been burnt the best thing to do is to throw it in the scraps; but if overheated it can be improved. Heat to a low red heat, and hammer lightly and cool off in salt water, while yet hot enough to be of a brown color. Repeat this a half a dozen times, and the steel will be greatly bettered. Of course, this is only in cases when a tool or something like it has been overheated which cannot be thrown away without loss. By this simple method I have restored tools overheated by ignorant smiths, and in some cases the owner would declare that it was "better than ever."

HOW TO DRESS AND HARDEN STONE HAMMERS

Care must be taken in heating stone hammers not to overheat them. Dress the hammer so that the edges are a little higher than the center, thus making a slight curve. A hammer dressed this way will cut better and stay sharp longer than if the face is level. Dress both ends before hardening, then harden face end first. Heat to a red heat, and cool off in cold water about one inch up, let the temper return to half an inch from the face, that is, draw the temper as much as you can without changing the temper at the face. There it should be as hard as you can make it. When heating the peen end keep a wet rag over the face to prevent it from becoming hot. This end should not be tempered quite as hard as the face.

HOW TO DRILL CHILLED CAST IRON

Chilled cast iron can be easily drilled if properly annealed, but it cannot be annealed simply by heating and slowly cooling. Heat the iron to a red heat and place it over the anvil in a level position; place a piece of brimstone just where the hole is to be drilled, and let it soak in. If it is a thick article place a piece on each side over the hole, as it will better penetrate and soften the iron. Next, heat it again until red, then bury it in the cinders, and let it cool slowly. To heat and anneal chilled iron is of no avail unless it is allowed to remain hot for hours. Chilled iron will, if heated and allowed to cool quick, retain its hardness. The only way to anneal is to let it remain in the fire for hours. Brimstone will help considerably, but even with that it is best to let cool as slowly as your time will admit.

HOW TO DRILL HARD STEEL

First, make your drill of good steel, oval in form, and a little heavier than usual on point, and temper as hard as it will without drawing the temper, the heat to be a low red cherry. Diluted muriatic acid is a good thing to roughen the surface with where you want the hole. Use kerosene instead of oil, or turpentine. The pressure on the drill should be steady so that it will cut right along as it is hard to start again if it stops cutting, but if it does, again use diluted muriatic acid. The hole should be cleaned after the use of the acid.

FACTS ABOUT STEEL

I have repeatedly warned against overheating steel. Don't heat too fast, for if it is a piece of a large dimension the outside corners will be burnt, while the bar is yet too cool inside to be worked. Don't let steel remain for any length of time in the fire at a high heat, for both steel and iron will then become brittle. This is supposed by some to be due to the formation of oxide disseminated through the mass of the metal, but many others believe that a more or less crystalline structure is set up under the influence of a softening heat, and is the sole cause of the diminution in strength and tenacity. The fiber of the steel is spoiled through overheating; this can, to some extent, be remedied by heavy forging if it is a heavy bar.

Steel is harder to weld than iron, because it contains less cinders and slag, which will produce a fusible fluid in iron that will make it weld without trouble. Steel contains from 2 to 25 per cent carbon, and varies in quality according to the per cent of carbon, and it is claimed that there are twenty different kinds of steel. To blacksmiths only a few kinds are known, and the sturdy smith discards both "physical tests and chemical analysis," and he thinks he knows just as much as do those who write volumes about these tests.

To weld tool steel, or steel of a high per cent of carbon, borax must be used freely to prevent burning and promote fusing. Steel with less carbon, or what smiths call "soft steel," "sleigh steel," should be welded with sand only. This soft steel stands a higher heat than the harder kinds.

Good tool steel will break easy when cold if it is cut into a little with a cold chisel all around, and the bar then placed with the cut over the hole in the anvil, the helper striking directly over the hole. If it is good steel it will break easy, and the broken ends are fine grain, of a light color. If it shows glistening or glittering qualities it is a bad sign.

Good steel will crumble under the hammer when white hot.

To test steel draw out to a sharp point, heat to a red heat, cool in salt water; if it cuts glass it is a steel of high hardening quality.

For armor piercing, frogs, tiles, safes, and crushing machinery, alloy steel is used. This steel contains chromium, manganese or nickel, which renders it intensely hard. Tungsten is another alloy that is used in iron-cutting tools, because it does not lose its hardness by friction. Smiths should know more about steel than they do, and we would have steel to suit every need. As it is now, any poor stuff is sent to the smith. The same can be said of iron. The American wrought iron is the poorest iron that ever got the name of iron, but there are thousands of smiths using this stuff with great difficulty without ever a word said as a protest against the manufacture of the rotten material.

We often get iron that is too poor to bend hot without breaking. Let us register a kick, and if that has no effect let us try to abolish the tariff, and there will be good iron manufactured in this country, or the Swedish and Norwegian iron will be used. But the result will be the same with iron as with the matches: the American manufactories will make good iron when they have to. We get iron and steel that is both "cold-shot and hot-shot." The former breaks easy when cold, the latter when hot. We have meat and wheat inspectors; where is the iron inspector? Farmers know enough to ask for protection, but blacksmiths will never say a word. They use the cold-shot or hot-shot iron, and when they have spent half a day in completing a little intricate work it breaks in their hands because of iron that is either cold or hot shot. Insist on good iron, and the steel will also be good. Deduct a little every year from the amount due your jobber for poor iron, and you can be sure if this is done by a few thousand smiths it will have effect.

HOW TO WELD CAST IRON

Strictly speaking, there is no such thing as welding cast iron. The best that can be done is to melt it together; but this is simply accidental work, and when done don't amount to anything. Still, I have never met a blacksmith yet who could not weld cast iron, but, at the same time, I have yet to meet the man that can do it; and I will give twenty-five dollars to the smith that will give me a receipt for welding cast-iron shoes that will be useful when welded. All receipts I have seen for this purpose are simply bosh.

Malleable iron is a different thing. Many smiths weld malleable iron and think it is cast iron. "The wish," in such a case, "is the father of the thought," but to weld malleable iron is not more difficult than to weld soft steel. Malleable iron when good, and steel when soft, are about the same thing. I would therefore advise smiths to spend no time in welding cast iron. Nothing will be gained even if you should succeed in sticking it enough to hang together. It will in most cases be useless, because it will not be of the same shape as before.

IF a pump handle is broken use rock salt and powdered glass as a welding compound. Stick the ends together in the fire. When they are about ready to melt tap lightly on one end while your helper holds the other end steady. In one case out of a hundred it will stick enough to hang together.

CASE HARDENING

Iron and steel may be case hardened with either of the following compounds: Prussiate of potash, sal-ammoniac of equal parts. Heat the iron red hot and sprinkle it with this compound, then heat again and sprinkle, and plunge it while yet hot in a bath of salt water.

Another: Cyanide of potassium; grind it into a fine powder and sprinkle over the iron while red hot, and plunge into a bath of salt water. This powder will coagulate if it is held against the fire so it gets warm Be careful with this powder, as it is a strong poison. It is the best thing that I have ever tried for case hardening iron. It will case harden the softest iron so that it cannot be touched with any tool. It is also good for plows, especially where it is hard to make a plow scour. The only objection is the price, as it costs more than prussiate of potash or other hardening compounds.

HOW TO HARDEN SPRINGS

Heat to a heat that will be discerned in the dark as a low red heat. Plunge into a bath of lukewarm water. Such a heat cannot be noticed in a light sunny day, but it is just the heat required.

Another way: Heat to a low red heat and bury the spring in cold sand. Another: Heat to a low red heat in the dark, and cool in oil.

TO MAKE STEEL AND IRON AS WHITE AS SILVER

Take 1 pound of ashes from white ash bark, dissolve in soft water. Heat your iron red, and cool in this solution, and the iron will turn white as silver.

TO MEND BROKEN SAWS

Silver, 15 parts; copper, 2 parts. These should be filed into powder and mixed. Now place your saw level with the broken ends tight up against each other; put a little of the mixture along the seam, and cover with powdered charcoal; with a spirit lamp and a blowpipe melt the mixture, then with the hammer set the joint smooth.

TO MEND A BAND SAW

If a band saw is broken file the ends bevel, and lap one end over the other far enough to take up one tooth; place the saw in such a position that the saw will be straight when mended; use silver, copper and brass; file into a fine powder; place this over the joint and cover with borax. Now heat two irons one inch square, or a pair of heavy tongs, and place one on each side of the joint, and when the powdered metal is melted have a pair of tongs ready to take hold over the joint with while it cools. File off and smooth the sides, not leaving the blade any thicker than in other places.

TO WRITE YOUR NAME ON STEEL

TAKE of nitric acid 4 ozs. muriatic acid, ½ oz. Mix together. Now cover the place you wish to write on with beeswax, the beeswax to be warm when applied. When it is cold, write your name with a sharp instrument. Be sure to write so that the steel is discernible in the name. Now apply the mixture with a feather, well filling each letter. Let the mixture remain about five minutes or more, according to the depth desired; then wash off the acid; water will stop the process of the same. When the wax is removed, the inscription is plain.

CHAPTER VIII

HOW TO PATCH A BOILER

By H. Moen, Machinist, Cresco, Iowa.

WHEN the leak or weak place in the boiler is found, take a ripping chisel and cut out all of the weak, thin and cracked parts. This done, make the patch. The patch must be large, not less than an inch lap on all sides, but if double rows of rivets are wanted the lap should be two inches on all sides. Bevel or scrape the patch on all edges to allow calking. The bolt holes should be about two inches apart and countersunk for patch bolts, if patch bolts are used. Next, drill two holes in the boiler shell, one on each side of the patch, and put in the bolts. These bolts should be put in to stay and hold the patch in position while the rest of the holes are drilled and bolted. When the bolts are all in, take your wrench and tighten the bolts one after the other, harder and harder, striking at the same time on the patch around its edges. At last strike light on the bolt heads when you tighten and draw the bolt until its head breaks off. These bolts are made for this purpose and in such a shape that the head will break at a high strain. This done use the calking iron all around the patch.

The patch should be put on the inside of the boiler, especially if on the bottom of a horizontal boiler. If the patch is put on the outside in this place the sediment or solid matter which the water contains will quickly fill up over the patch and there is danger of overheating the boiler and an explosion may follow.

HOW TO PUT IN FLUES

The tools necessary to retube an old boiler are, first, a good expander of the proper size; a roller expander preferred; a crow foot or calking iron, made from good tool steel. A cutting-off tool can be made to do very

TUBE TO BE WELDED

good service, in the following manner: Take a piece of steel, say ½ x 1¼, about ten inches long. Draw one end out to a sharp point and bend to a right angle of a length just enough to let it pass inside of the flue to be cut. A gas pipe can be used for a handle. In cutting the flues set this tool just inside the flue sheet and press down on the handle. If this tool is properly made it cuts the old flues out with ease. After both ends have been cut the flues will come out.

MODERN BLACKSMITHING

Next, cut the tubes about ⅜ of an inch longer than the flue sheet. After the tubes are cut the proper length, and placed in the boiler, expand the same in both ends with a flue expander. After the flues are expanded until they fit the holes solid, turn them over with the peen of a hammer to make them bell shaped. Now take a crows-foot, or calking tool, and turn the

TUBE EXPANDER

ends in a uniform head and tight all around. If the flues should leak, and there is water on the boiler take a boiler expander and tighten them up. But never attempt to tighten a flue with the hammer if there is water on the boiler.

HOW TO WELD FLUES

In welding flues or putting new tips on old flues, you must find out how far the old tubes are damaged, and cut that part off. Next clean the scales off in a tumbling box; if you have none, with an old rasp.

Now take a piece of tubing the size of the old, and scarf the ends down thin, the new tube to go over the old and drive them together. In welding a rest can be made in the forge to push the tube against while welding, to prevent the pieces from pulling apart. A three-eighths rod, with thread on one end and a head on the other, run through the flue will be found handy for holding the pipes or flues together. In welding these together don't take them out of the fire and strike with a hammer, but take a rod ⅜-inch round, and bend one end to serve as a hammer. Strike with this hammer lightly over the lap, at the same time turning the flue around in the fire. Use borax to prevent the flue from scaling and burning.

FOAMING IN BOILERS

There are many reasons for foaming in boilers, but the chief reason is dirty water. In some cases it is imperfect construction of boiler, such as insufficient room for the steam and a too small steam pipe or dome. When a boiler is large enough for the steam and clean water is used there is no danger of foaming. When more water is evaporated than there is steam room or heating surface for, then the boiler will foam. When a boiler is overworked more steam than its capacity will admit is required, and the engine is run at a high speed, the steam will carry with it more water than usual.

When a boiler foams shut the throttle partly to check the outflow of steam and lessen the suction of water, because the water is sucked up and follows the sides of the dome up.

If the steam pipe in the dome sticks through the flange a few inches the water will not escape so easy. A boiler that is inclined to foam should not be filled too full with dirty water; if it is it is best to blow off a little. Foul water can be cleaned by different methods before it enters the boiler, so as to prevent foaming and scaling.

BLOWING OUT THE BOILER

A boiler should not be blowed out under a high steam pressure, because the change is so sudden that it has a tendency to contract the iron, and if repeated often the boiler will leak. If it is done when there is brickwork around the boiler and the same is hot it will in a short time ruin the boiler. In such a case the boiler should not be blowed out for hours after you have ceased firing.

DESIGN AND TREATMENT OF HOISTING HOOKS

The construction and treatment of a hook intended for a crane or hoist involves a problem deserving of careful consideration. This fact has impressed itself upon the writer from his experience in lawsuits arising from damages caused by the failure of a defective hook.

To make a hook safe for the purpose for which it is intended two important requisites must be fulfilled—
 1. Correct design.
 2. Proper treatment.

The design is guided to a large extent by the service which the hook is to perform; that is, whether the hook will be subjected to high or low, frequent or rare stresses, and whether the hook is likely to be loaded above its normal capacity.

Hooks on small cranes and hoists, of about 2 to 3 tons' capacity, may be loaded to full capacity several times every day; while hooks on cranes of 50 tons' capacity may carry full load at remote intervals only. Due to the repeated stresses on small hooks, occurring, possibly, under extreme temperatures, the hook may become fatigued and liable to break. It is, therefore, advisable to keep the stresses low in comparatively small hooks to provide a factor of safety.

This precaution can readily be observed with hooks for small loads, as the size would scarcely render them unwieldy. But as the size of the hook increases it becomes necessary to increase the stress; that is, the load per square inch or cross section, in order to avoid the construction of a clumsy hook. A high stress is permissible with high loads because they are applied to the hook less frequently than in the case of small hooks and light loads. We may consider a stress of 15,000 lbs. per sq. in. as safe for a 50-ton hook as a stress of 10,000 lbs. per sq. in. on a 10-ton hook.

The material for a hook may be ordinary steel, cast steel or wrought iron; depending on the load the hook is to carry. For small loads where a hook of ample size, yet not bulky, can be constructed, cast steel may be used. But for heavy loads a ductile material, having practically the same elastic limit for compression and tension, should be selected.

Mr. F. A. Waldron, for many years connected with the manufacturing of hooks in the works of the Yale & Towne Mfg. Co., made careful observations with different materials, and his conclusion is, that the only reliable material for hooks is a high grade puddled iron. A steel hook may carry a load from 25 to 50% greater than the wrought iron hook, but it is not reliable. This fact will be borne out more clearly by

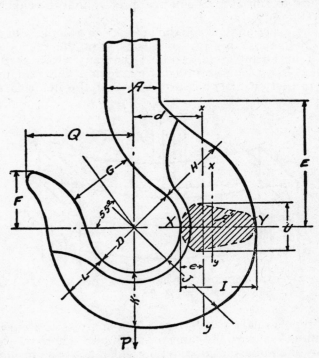

FIG. 1—THE CORRECTLY PROPORTIONED HOOK IS BASED UPON DEFINITE FORMULAE

the results of tests made by the writer and given further in this article.

The design of a hook should be based on formulae deduced from practice with successful hooks, rather than to depend on theoretical computations. In the latter method, conditions are assumed which are hardly ever realized in actual practice. It is, therefore, absurd to aim at mathematical precision at the expense of reliability.

The exact analysis of the stresses in a hook is based on the theory of curved beams. In the theory of straight beams it is assumed that any cross section which is a plane section before flexure will remain a plane section after flexure, and that the deformation

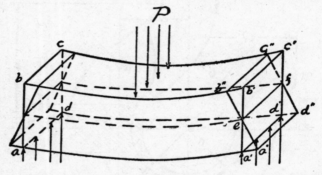

FIG. 2—THE HOOK CONSIDERED A BEAM WITH THE LOAD AT P

is proportional to the stress. The analysis of a curved beam is based on the same assumption. There is, however, one important distinction which has been brought out by recent investigations.

Consider a straight beam loaded transversely with a load P, as shown by Fig. 2. Originally, the fibers between the cross sections a-b-c-d and a'-b'-c'-d' were of the same length. When loaded, the fibers in the strip b-c, b'-c' are subjected to compression; and the fibers in the strip a-d, a'-d' are under tension; consequently, the upper fibers will shorten and the lower fibers will lengthen. Somewhere between b-c, b'-c' and a-d, a'-d', there must exist a layer of fibers which have neither shortened nor elongated. The intersection of this layer with the section a''-b''-c''-d'' (the position of a'-b'-c'-d' when the beam is loaded) is a straight line (e, f) called the "neutral axis", which for straight beams coincides with the gravity axis of the cross section.

In a curved beam the neutral axis, x-y, Fig. 1, does not coincide with the gravity axis, x'-y', Fig. 1, but falls somewhere between the gravity axis and the tension side of the hook. This is due to the fact that the fibers on the convex side of the hook are longer than those on the concave side, and therefore require less stress than the shorter fibers for the same amount of deformation.

The application of the theory of curved beams is somewhat complicated for practical purposes, and a simpler form can be used; provided that care is taken in assigning the limits of stress. Referring to Fig. 1, assume the beam with a load P. The most dangerous section is, evidently, along X-Y; it is acted upon by a direct tension stress (f' $= P/A'$) and a flexure stress (f'') due to the bending moment (P $\times$ a); the combined stress is the sum of f' and f''. Let f represent the combined stress;

$$\text{then } f = \frac{P}{A'} + \frac{P\,d\,e}{I}$$

in which A' is the area of the cross section, e the distance of the neutral axis from the tension side, and I the moment of inertia of the section about the neutral axis.

If the material has, practically, the same elastic limit

for compression and tension, the neutral axis needs not be far from the gravity axis; otherwise, it is advisable to distribute the metal more toward the tension side. The most reliable data on the construction of hooks, resulting from extensive experimental and mathematical investigations, is that given by Mr. Henry R. Towne in his treatise on cranes.

The following formulae for determining the dimensions of the various portions will give the greatest resistance to spreading and rupture that the original bar will permit. Referring again to Fig. 1:

$$D = .5 \Delta + 1.25$$
$$H = 1.08 A$$
$$L = 1.05 A$$
$$U = .886 A$$
$$G = .75 D$$
$$F = .33 \Delta + .85$$
$$I = 1.33 A$$
$$J = 1.20 A$$
$$K = 1.13 A$$
$$E = .64 \Delta + 1.6$$
$$Q = .64 \Delta + 1.6$$

In the above formulae the dimensions are in inches; Δ is the load in tons of 2,000 lbs.

Having discussed, to some extent, the design of the hook, we may now consider the care which should attend the making of a hook. As already mentioned, the writer on several occasions gave expert testimony on lawsuits before the court arising from injuries incurred and damages caused by the failure of defective hooks.

To investigate the subject thoroughly, a series of careful experiments were made to determine how a hook should be made to insure reasonable safety and facilitate the location of responsibility in case of failure.

Hooks may become hardened in course of time when exposed to extreme temperatures. Repeated stresses combined with extreme temperatures will change the molecular structure and, to some extent, the physical properties of the metal.

The fact that chain links and hooks break more often in cold weather, suggests the advisability of annealing chains and hooks at suitable intervals, to refine the grain which may have crystallized. A crystallized grain is always a weak grain and is especially objectionable where a sudden or "shock" load is likely to be applied.

In conclusion it may be remarked that the courts do not consider the correct size of the stock as the determining factor when locating the responsibility in case of accident. If it can be shown that the hook was too hard, overheated or too high in carbon, or there was a flaw in workmanship, there is undisputed evidence of negligence which constitutes sufficient ground for suit to recover damages.

There is a simple method for avoiding accidents by failure of hooks. If you buy or make your hooks, see to it that they contain the proper amount of carbon, and anneal before using; continue annealing, at least once a year, if the hooks are subjected to varying and extreme temperatures.

CHAPTER IX

THE HORSE

THE horse must have been one of the first animals subjected to the use of man, but there is no record made of it before the time of Joseph, during the great famine in Egypt, when Joseph exchanged bread for horses. During the exodus horses were used more extensively, and in consequent wars we find the horse used especially by great men and heroes. This noble animal has always been held in high esteem by civilized people. In wars and journeys and exploits, as well as for transports, the horse is of immeasurable value. No people cared for and loved this animal as did the Arabs. The care and breeding of horses was their main occupation, therefore their horses were noted for intelligence, high speed and endurance. The English and American thoroughbred has an infusion of blood of the Arab horse, which has set the price on these animals. Pedigrees were first established by the Arabs.

Each country has its own breed of horses. Horses

of a cold climate are smaller in size, as also are the horses of the tropics. The best horses are found in the temperate zone. In Germany the horses are large, well formed and strong. Norway and Sweden have a

FIG. 3.

race of little horses, and not until a few years ago did the people of these countries know anything about pedigrees; their horses are spirited and stronger in porportion to the size than any other race of horses. In Sweden and Norway the farmer, with wife and children, will walk many miles Sunday to church.

while the horses roam in the pasture or stand in the stable. Some farmers will not hire out their team for money. The horses of these countries are better taken

FIG. 51.—TOE AND SIDE WEIGHT AND PLAIN RACING PLATES, AS MANUFACTURED BY BRYDEN HORSE SHOE CO.

care of than anywhere else, of course with the exception of American race horses.

HORSE-SHOEING

The horse in a wild state needs no shoes, the wear and tear that the feet are subjected to while the horse is hunting for his food in a wild country on soft meadows, is just right to keep the hoofs down in a normal condition. But when the horse is in bondage and must serve as a burden-carrying animal, traveling on hard roads or paved streets, the horse must be shod to prevent a foot wear which nature cannot recuperate. Horseshoes were first made of iron in 480 A. D. Before that time, and even after, horseshoes have been made of leather and other materials.

FIG. 52.—TOE AND SIDE WEIGHT AND PLAIN RACING PLATES, AS MANUFACTURED BY BRYDEN HORSE SHOE CO.

ANATOMY

It is necessary in order to be a successful horse-shoer to know something about the anatomical construction of the feet and legs of the horse. Of course, any little boy can learn the names of the bones and tendons in a horse's foot in an hour, but this does not make a horse-shoer out of him. No board of examiners should

allow any horse-shoer to pass an examination **merely**
because he can answer the questions put to him **in**
regard to the anatomy of the horse, for as I have **said**
before, these names are easily learned, but **practical**

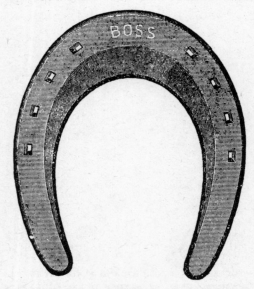

FIG. 53.—TOE AND SIDE WEIGHT AND PLAIN RACING **PLATES, AS**
MANUFACTURED BY BRYDEN HORSE SHOE CO.

horse-shoeing is not learned in hours; it takes years **of**
study and practice.

It is not my intention to treat on this subject. I could
not; first, because there is not room for such a dis-
course; second, there are numerous books on the subject better than I could write, available to **every**

horse-shoer. I shall only give a few names of such parts of the anatomy as is essential to know. What the horse-shoer wants to know is the parts of the foot connected with the hoof, as his work is confined solely to the foot.

FIG. 54.—TOE AND SIDE WEIGHT AND PLAIN RACING PLATES, AS MANUFACTURED BY BRYDEN HORSE SHOE CO.

THE WALL

The wall or crust is the horny sheath incasing the end of the foot, in the front and on the sides from the coronet to the ground. It is through this crust the nail is driven, and it is upon this crust the shoe rests. In front it is deepest, towards the quarter and heel it becomes thinner. It is of equal thickness from the upper end to the ground (from top to bottom). The white corored wall is the poorest, while the iron colored wall is the toughest. The growth of the wall is different at different ages. It grows more in a young horse and colt than in an old horse; in a healthy foot and soft, than in a diseased foot and hard. In a young horse the hoof will grow about three inches in a year and even more, while it grows less in an old horse. The wall is fibrous, the fibers going parallel to each other from the coronet to the ground.

THE SOLE

The horny sole is the bottom of the foot. This sole is fibrous like the wall. The sole is thickest at the border, where it connects with the wall, and thinnest at the center. The sole when in a healthy condition scales off in flakes. This scale is a guide to the farrier whereby he can tell how much to pare off. There are different opinions in regard to the paring of the sole, but that is unnecessary, for nature will tell how much to cut off in a healthy foot. In a diseased foot it is different; then the horse-shoer must use his own good judgment. It is, however, in very few cases that the shoer needs to do more than just clean the sole. Nature does the scaling off, or paring business, better than any farrier.

THE FROG

The frog is situated at the heel and back part of the hoof, within the bars; the point extending towards the center of the sole, its base filling up the space left between the inflection of the wall. This body is also fibrous. The frog is very elastic and is evidently designed for contact with the ground, and for the prevention of jars injurious to the limbs.

CORONET

Coronet is the name of the upper margin of the foot, the place where the hair ceases and the horny hoof begins.

THE QUARTER

The quarter means a place at the bottom of the wall, say, about one-third the length from the heel towards the toe.

THE BARS

By the bars we mean the horny walls on each side of the frog, commencing at the heel of the wall and extending towards the point of the frog.

Any blacksmith or horse-shoer desiring to study more thoroughly the anatomy of the horse should procure a book treating on this subject.

HOW TO MAKE THE SHOE

It is only in exceptional cases that the shoer turns or makes a shoe. The shoes are now already shaped, creased and partly punched, so all that is needed is to weld on the toe calk and shape the heel calks.

Heat the shoe at the toe first, and when hot bend the heels together a little. This is done because the shoes will spread when the toe calk is welded on, and the shoe should not be too wide on the toe, as is mostly the case. If the shoe is narrow at the toe it is easier to fit the same to the foot and get the shoe to fill out on the toe. Many smiths cut too much off from the toe. Before the toe calk is driven onto the shoe bend it a little so as to give it the same curve the shoe has, and the corners of the calk will not stick out over the edge of the shoe. Now place the shoe in the fire, calk up. Heat to a good low welding heat, and use sand for welding compound. Don't take the shoe out of the fire to dip it in the sand, as most shoers do, for you will then cool it off by digging in the cold sand, of which you will get too much on the inner side of the calk. The same will, if allowed to stay, make the calk look rough. You will also have to make a new place for the shoe in the fire, which will take up a good deal of time, as the new place is not at once so hot as the place from which the shoe was taken; besides this, you might tear the calk off and lose it. When hot give a couple of good blows on the calk and then draw it out. Don't hold the heels of the shoe too close to the anvil when you draw out the calk, for if you do the calk will stand under, and it should be at a right angle with the shoe. Do not draw it out too long, as is mostly done. Punch the hole from the upper side first. Many first-class horse-shoers punch only from that side, while most shoers punch from both sides.

There is no need of heating the shoe for punching the holes. Punch the holes next to the heel first, for if you punch the holes next to the toe when the shoe is hot, the punch will be hot, upset and bent. If it is a large shoe, punch only two holes on each side for the toe calk heat. These holes to be the holes next to the toe when the shoe is hot, and then punch the other two when you draw out the heel calks, and the shoe is hot at the heel. The heel calks should be as short as you can make them; and so should the toe calks. I know but a few horse-shoers that are able to weld on a toe calk good. The reason for their inability is lack of experience in general blacksmithing. Most shoers know not how to make a fire to weld in. They are too stingy about the coal; try to weld in dirt and cinders, with a low fire, the shoe almost touching the tuyer iron. I advise all horse-shoers to read my article about the fire.

I have made a hammer specially for horse-shoeing with a peen different from other hammers. With this hammer the beginner will have no trouble in drawing out the calks. See Figure 8, No. 8. The hammers as now used by most smiths are short and clumsy; they interfere too much with the air, and give a bump instead of a sharp cutting blow that will stick to the calk.

The shoe should be so shaped at the heel as to give plenty of room for the frog; the heels to be spread out as wide as possible. This is important, for if the shoe is wide between the heels the horse will stand more firm, and it will be to him a comfortable shoe. The shoe should not be wider between the calks at the expense of same, as is done by some shoers, for this is only a half calk, and the heel is no wider. The shoe should not be fitted to the foot when hot, as it will injure the hoof if it is burned to the foot.

HOW TO PREPARE THE FOOT FOR THE SHOE

The foot should be level, no matter what the fault is with the horse. The hoof should not be cut down more than the loose scales will allow. In a healthy condition this scale is a guide. When the foot is diseased it is different, and the shoer must use his own judgment.

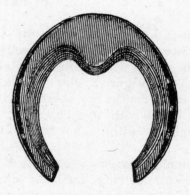

The frog never grows too large. It should never be trimmed more than just to remove any loose scales.

The frog in its functions is very important to the well-being of the foot. In the unshod foot it projects beyond the level of the sole, always in contact with the ground; it obviates concussion; supports the tendons; prevents falls and contraction. The bars are also of importance, bracing the hoof, and should never be cut down as has been the practice for centuries by ignorant horse-shoers.

FORGING

Forging or overreaching is a bad habit, and a horse with this fault is now very valuable. This habit can be overcome by shoeing; but it will not be done by making the shoes short on the heel in front and short in the toe behind. Never try this foolish method.

To overcome forging the shoer should know what forging is. It is this: The horse breaks over with his

hind feet quicker than he breaks over with the front feet; in other words, he has more action behind than in front, and the result is that the hind feet strike the front feet before they can get out of the way, often cutting the quarters badly, giving rise to quarter cracks and horny patches over the heel.

Some writers make a difference between forging and overreaching, but the cause of the trouble is the same —too much action behind in proportion to the front; and the remedy is the same—retard the action behind, increase it in front. There are different ideas about the remedy for this fault.

One method is to shoe heavy forward and light behind, but this is in my judgment a poor idea, although it might help in some cases. Another way is to shoe

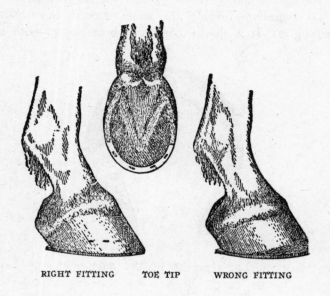

RIGHT FITTING   TOE TIP   WRONG FITTING

with side weight on the outer side behind, but it is not safe, because it is difficult to get a horse to throw the foot out to one side enough so as to pass by the front foot except in a high trot.

The best way to shoe a forger or overreaching horse is to make a shoe for front of medium heft, not longer than just what is needed. The toe calk should be at the inner web of the shoe, or no toe calk at all, or, toe weight, to make the horse reach farther.

It will sometimes be found that the hind foot is shorter than the front foot. To find this out, measure from the coronet to the end of the toe. The shorter the foot the quicker it breaks over. If it is found that the hind foot is shorter than the front foot, then the shoe should be made so that it will make up for this. Let the shoe stick out on the toe enough to make the foot of equal length with the front foot. It is well in any case of forging to make the hind shoe longer on the toe. If the hind shoe is back on the foot, as is often done, it will only make the horse forge all the more, for it will increase action behind, the horse breaks over quicker, and strikes the front foot before it is out of the way. Set the shoe forward as far as possible, and make long heels. The longer the shoe is behind the longer it takes to raise the foot and break over.

Clack forging is meant by the habit of clacking the hind and fore shoes together. This kind of forging is not serious or harmful; it will only tend to wear off the toe of the hind foot and annoy the driver, possibly a little fatiguing to the horse.

The position of the feet at the time of the clack is different from that it is supposed to be. The toe of the hind feet is generally worn off, while no mark is made on the front feet. From this you will understand that the hind feet never touch the heel of the front feet, but the shoe. Just at the moment the fore foot is raised up enough on the heel to give room for the hind foot to wedge in under it the hind foot comes flying under the fore foot, and the toe of the hind foot

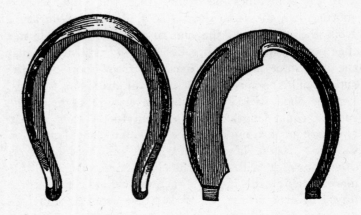

strikes the web of the toe on the front foot. This is the reason no mark is seen on the front foot, while the hind foot is badly worn off.

INTERFERING

Interfering is a bad fault in a horse. It is the effect of a variety of causes. In interfering the horse brushes the foot going forward against the other foot. Some horses strike the knee, others above it, the shin or coronet, but in most cases the fetlock.

Colts seldom interfere before they are shod, but then they sometimes interfere because the shoes are too heavy. This trouble disappears as soon as the colt is accustomed to carrying the shoes. Weakness is the most common cause. Malformation of the fetlock is another cause. The turning in or out of the toes, giving a swinging motion to the feet, is also conducive to interfering.

The first thing to do is to apply a boot to the place that is brushed. Next, proceed to remove the cause by shoeing, or by feeding and rest in cases of weakness. Nothing is better than flesh to spread the legs

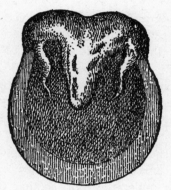

NATURAL FOOT

with. Some old horse-shoers in shoeing for interfering will turn the feet so as to turn the fetlock out. This is done by paring down the outside and leaving the inside strong. This is a bad way of shoeing for interfering, as it might ruin the horse. The foot should be leveled as level as it is possible. The inner side of the hoof should be scant; instead of being curved it should be almost straight, as the horse generally strikes with the side of the hoof or quarter. This is done to make a side-weight shoe, the side weight not to reach over the center of the shoe, but to be only on one side. Put the shoe on with the weight on the outer side. If the horse still interferes, give more side weight to the shoe, and make the heel on the outer side about one and one-quarter inch longer than the inside heel; give it an outward turn. This heel will prevent the horse from turning the heel in the way of the other foot when it goes by, so as not to strike the fetlock.

Properly made and applied, side weight will stop interfering almost every time. If the side weight is heavy enough it will throw the foot out, and the trouble is overcome.

There are only a few horse-shoers that have any practical experience in making side-weight shoes, which we understand from the articles in our trade journals.

Some horse-shoers in shoeing to stop interfering will make common shoes shorter than they ought to be and set them far in under the foot, so that the hoof on the inner side will stick out over the shoe a quarter of an inch. These they don't rasp off, and everybody knows that the hoof adheres to and rubs harder against the leg than the hard smooth shoe. But, foolish as it is, such shoers stick to their foolish ideas. I call all such fads faith cures.

The rule is to have the side weight on the outer side, while the exception is to have the side weight on the inner side of the foot. For old and poor horses ground feed and rest is better than any kind of shoes. It will give more strength and more flesh to spread the legs.

CHAPTER X

HOW TO SHOE A KNEESPRUNG OR KNUCKLER

KNEESPRUNG is the result of disease that sometimes is brought about by bad shoeing. In a healthy leg the center of gravity is down through the center of the leg and out at the heels. This is changed in a case of kneesprung legs, giving the legs a bowed appearance. This trouble always comes on gradually; in some cases it will stop and never get worse, while in others it will keep on until it renders the horse useless. A horse with straight legs will sleep standing, but a knuckler cannot; he will fall as soon as he goes to sleep, on account of the center of gravity being thrown on a line forward of the suspensory ligaments. The cause of this trouble is sprain or injury to the back tendons of the legs; soreness of the feet, shins or joints. In old cases nothing can be done but just to relieve the strain a little by shoeing with a long shoe and high heel calks, with no toe calk. In cases not more than three months old clip the hair off the back tendons when there is any soreness, and shower them with cold water

several times a day for a week or two, and then turn the horse out for a long run in the pasture.

CONTRACTION

Contraction is in itself no original disease, except in a few cases. It is mostly the effect of some disease.

FOOT PREPARED FOR CHARTIER TIP   FOOT SHOD WITH CHARTIER TIP

Contraction follows sprains of the tendons, corns, founder and navicular disease. When contraction is the result of a long-standing disease of the foot or leg it will be in only one of the feet, because the horse will rest the affected leg and stand most of the time on the healthy leg; thus the healthy foot receives more pressure than the diseased, and is spread out more; the foot becomes much uneven—they don't look like mates. This kind of contraction is generally the result of some chronic disease, but in most cases contraction is the result of shoeing and artificial living. Before the colt is shod his hoofs are large and open-heeled, the quarters are spread out wide, and the foot on the under side is shaped like a saucer. The reason of the colt's foot being so large is that he has been running on the green and moist turf, without shoes, and the feet have in walking in mud and dampness gathered so much moisture that they are growing and spreading at every

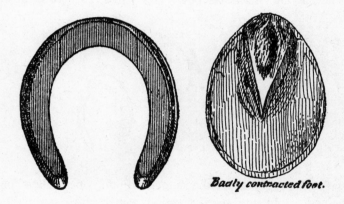

Badly contracted foot.

step. This is changed when the colt is shod and put on hard roads, or taken from the pasture and put on hard floors where the feet become hard and dried up. A strong high heeled foot is predisposed to contraction, while a low heeled flat foot is seldom afflicted with this trouble.

When contraction comes from bad shoeing or from standing on hard floors, pull the shoes off, pare down the foot as much as you can, leaving the frog as large as it is. Rub in some hoof ointment once a day at the coronet and quarters, and turn the horse out in a wet pasture. But if the horse must be used on the road, proceed to shoe as follows: First, ascertain if the frog is hard or soft. If soft, put on a bar shoe with open bar. I have invented a shoe for this purpose. See Fig. 2, No. 1. The idea of shoeing with an endless bar shoe is wrong. In most cases contraction is brought on by letting the shoes stay on too long, whereby the hoof has been compelled to grow down

with the shape of the shoe. If an open shoe has helped to bring on contraction, much more so will a bar shoe, which will tie the hoof to the shoe with no chance of spreading, no matter what frog pressure is put on. Make the shoe as light as you can, with very low or no calks; let the bar rest against the frog; keep the hoofs moist with hoof ointment; use an open bar shoe.

Make a low box and fill it with wet manure, mud or clay, and let the horse stand in it when convenient, to soften the hoofs. Spread the shoe a little every week to help the hoofs out, or the shoes will prevent what the frog pressure aims to do, but this spreading must be done with care. If the frog is dried up and hard, don't put on a bar shoe, as it will do more harm than good. In such a case make a common shoe with low or no calks; make holes in it as far back as you can nail; spread them with care a little every week. Let the horse stand in a box with mud or manure, even warm water, for a few hours at a time, and keep the hoofs moist with hoof ointment. In either case do not let the shoe stay on longer than four weeks at a time. In addition to the above pack the feet with some wet packing, or a sponge can be applied to the feet and held in position by some of the many inventions for this purpose.

No man can comprehend how much a horse suffers from contraction when his feet are hoof-bound and pressed together as if they were in a vise. The pain from a pair of hard and tight boots on a man are nothing compared to the agony endured by this noble and silent sufferer. It must be remembered that there is no such a thing as shoeing for contraction. Contraction is brought on by artificial living and shoeing. A bar shoe for contraction is the most foolish thing to imagine. The pressure intended on the frog is a dead pressure, and in a few days it will settle itself so that there is no pressure at all. If a bar shoe is to be used it must be an open bar shoe like the one referred to. This shoe will give a live pressure, and if made of stell will spring up against the frog at every step and it can be spread. I will say, however, that I don't recommend spreading, for it will part if not done with

154 MODERN BLACKSMITHING

care. It is better to drive the shoe on with only **four or** five nails, and set them over often. Contraction **never**

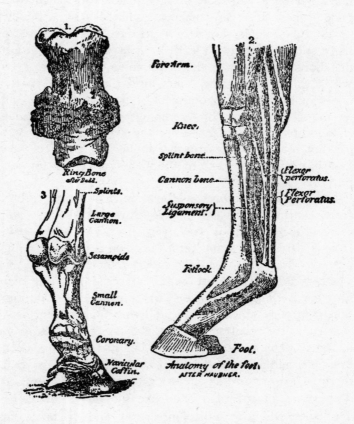

affects the hind feet because of the moisture they receive. This should suggest to ever shoer that moisture is better than shoes.

CORNS

Corns are very common to horses' feet, a majority of all cases of lameness is due to this trouble.

Corns are the result of shoes being allowed to stay on too long. The shoe, in such a case grows under the foot and presses on the sole and corns are formed. Even pressure of the shoe and sometimes too heavy bearing on the heel causes corns. Gravel wedging in under the shoe or between the bar and the wall is sometimes the cause of corns. Leaving the heel and quarters too high, whereby they will bend under and press against the sole, is another cause of corns.

The seat of corns is generally in the sole of the foot at the quarter or heels between the bar and the wall, at the angle made by the wall and bar.

Anything that will bruise the underlying and sensitive membrane of the sole will produce corn. This bruise gives rise to soreness, the sole becomes blood colored and reddish; if bad it might break out, either at the bottom or the junction of the hoof and hair or coronet, forming a quittor.

Cut out the corn or red sole clear down. If the corn is the result of contraction pare down the hoof and sole, put the foot into linseed poultice that is warm, for twenty-four hours, then renew it. If the corn is deep, be sure to cut down enough to let the matter out. It is a good thing to pour into the hole hot pine tar. In shoeing the bearing should be taken off the quarter or from the wall over the corn by rasping it down so that it will not touch the shoe. A bar shoe is a good thing as it will not spring as much as to come in contact with the hoof over the corn. Give very little frog pressure. An open shoe can be used and in

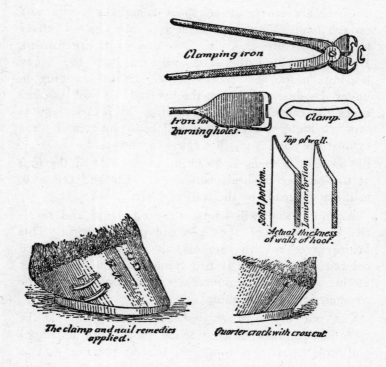

such a case there should be no calk at the heel. A calk should be welded on directly over the corn and the shoe will not spring up against the wall.

QUARTER AND SAND CRACKS

Quarter and sand cracks are cracks in the hoof, usually running lengthwise of the fibers, but sometimes they will be running across the fiber for an inch

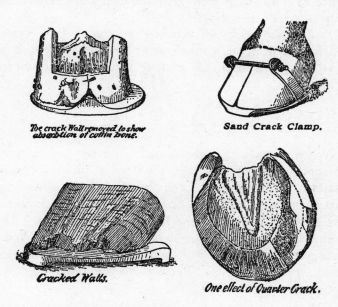

Toe crack Wall removed to show absorption of coffin bone.

Sand Crack Clamp.

Cracked Walls.

One effect of Quarter Crack.

or more. Quarter cracks are cracks mostly on the inside of the hoof, because that side is thinner and weaker than the outside. The cause of it is a hard and brittle hoof with no elasticity, brought on by poor assimilation and a want of good nutrition to the hoof. Hot, sandy or hard roads are also conducive to these cracks. What to do: If the horse is shod remove the shoes, and cut off the wall of the quarter to take off the bearing on both sides of the crack. If the crack goes up to the coronet and is deep, cut off both sides of the crack the whole length. About one inch below the coronet, cut a deep cut clear through either with a knife or hot sharp iron across the crack. This will help to start a new hoof.

If the flesh sticks up between the cracks, let a veterinarian burn it off. In shoeing for this trouble, it is best to use a bar shoe (endless) and shoe the horse often.

SEEDY TOE

When shoes with a clip or a cap on the toe are used it sometimes happens that the toe is bruised and it starts a dry rot extending up between the wall and the laminæ. Remove the shoe, pare away the hoof at the toe so as to take away the bearing from the toe. Any white or meaty substance should be picked out. Apply hot pine tar into the hole, and dip a little wad of tow in the hole to fill up. Replace the shoe, but don't let the clip touch the wall.

PRICKING

Pricking often happens in shoeing from a nail running into the quick, but the horse is often pricked by stepping on a nail or anything that will penetrate the sole and run into the quick. If the horse is pricked by shoeing pull off the shoes and examine each nail, the nail which has gone into the quick is wet and of a blue color.

If it is a bad case the sole or wall must be cut down to let the matter out and the foot put into a boot of linseed poultice. In milder cases a little pine tar put into the hole will be enough.

STIFLED

Mistakes are often made by inexperienced men and horse-shoers when a case of this kind is to be treated, and I would advise every horse-shoer to call in a veterinarian when he gets a case of this kind. Cramps of the muscles cf the thighs are sometimes taken for stifle.

When stifle appears in an old horse, three ounces of lead through his brain is the best, but for a young horse a cruel method of shoeing might be tried. Make a shoe with heels three inches high, or a shoe with cross bands as shown in illustration, Figure 8, No. 2, for stifle shoe. This shoe must be placed on the well foot. The idea is to have the horse stand on the stifled leg until the muscles and cords are relaxed.

STRING HALT

String halt or spring halt is a kind of affection of the hind legs, occasioning a sudden jerk of the legs upward towards the belly. Sometimes only one leg is affected.

160 MODERN BLACKSMITHING

In some cases it is milder, in others more severe. In some cases it is difficult to start the horse. He will jerk up on one leg and then on the other, but when started will go along all right.

For this fault there is no cure because it is a nervous affection. If there is any local disorder it is best to treat this, as it might alleviate the jerk. For the jerk itself bathe the hind quarters once a day with cold water. If this don't help try warm water, once a day for two weeks. Rub the quarters dry after bathing.

HOW TO SHOE A KICKING HORSE

Many devices are now gotten up for shoeing kicking horses. It is no use for a man to wrestle with a horse, and every horse-shoer should try to find out the best way to handle vicious horses.

One simple way, which will answer in most cases, is to put a twist on one of the horse's lips or on one ear. To make a twist, take a piece of broom handle two feet long, bore a half-inch hole in one end and put a piece of a clothes line through so as to make a loop six inches in diameter.

Another way: Make a leather strap with a ring in, put this strap around the foot of the horse; in the ring of the strap tie a rope. Now braid or tie a ring in the horse's tail and run the rope through this ring and back through the ring in the strap, then pull the foot up. See Fig. 16. The front foot can be held up by this device also, by simply buckling the strap to the foot and throwing the strap up over the neck of the horse.

Shoeing stalls are also used, but they are yet too expensive for small shops.

No horse-shoer should lose his temper in handling a nervous horse and abuse the animal; for, in nine cases out of ten, will hard treatment make the horse worse, and many horse owners would rather be hit themselves than to have anybody hit their horse.

EASY POSITION FOR FINISHING

Don't curse. Be cool, use a little patience, and you will, in most cases, succeed. To a nervous horse you should talk gently, as you would to a scared child. The horse is the noblest and most useful animal to man, but is often maltreated and abused. Amongst our dumb friends, the horse is the best, but few recognize this fact.

HOW TO SHOE A TROTTER

In shoeing a trotter it is no use to follow a certain rule for the angle, because the angle must vary a little in proportion to the different shape of the horse's foot.

Every owner of a trotter will test the speed by having shoes in different shapes and sizes, as well as having the feet trimmed at different angles, and when the angle is found that will give the best results the owner will keep a record of the same and give the horseshoer directions and points in each case.

The average weight of a horse-shoe should be eight ounces. Remember this is for a trotter. Make the shoe fit to the edges of the wall so that there will be no rasping done on the outside. In farm and draft horses this is impossible, as there is hardly a foot of such a uniform shape but what some has to be rasped off.

Use No. 4 nails, or No. 5.

Don't rasp under the clinches of the nails.

Make the shoes the shape of No. 1, Figure 8.

HOW TO SHOE A HORSE WITH POOR OR BRITTLE HOOFS

Sometimes it is difficult to shoe so as to make the shoe stay on on account of poor and brittle hoofs. In such a case the shoe should be fitted snug. Make a shoe with a toe clip.

HOW TO SHOE A WEAK-HEELED HORSE.

In weak heels the hoof is found to be low and thin from the quarters back. The balls are soft and tender. The shoes should not touch the hoof from the quarters back to the heels. An endless bar shoe is often the best thing for this trouble, giving some frog pressure to help relieve the pressure against the heels.

FOUNDER

Founder is a disease manifested by fever in the feet in different degrees from a simple congestion to a severe inflammation. It is mostly exhibited in the fore feet, being uncommon in the hind feet. The reason for this is the harder pressure, a much greater amount of weight coming on the front feet, the strain and pressure on the soft tissues heavier. The disease is either acute or chronic, in one foot or both. When both feet are diseased the horse will put both feet forward and rest upon the heels so as to relieve the pressure of the foot. If only one foot is affected that foot is put forward and sometimes kept in continual motion, indicating severe pain. The foot is hot, especially around the coronary band. The disease, if not checked, will render the horse useless. When such a horse is brought to you for shoeing it would be best to send him to a veterinarian.

How to shoe: Let the horse stand in a warm mud puddle for six hours, then put on rubber pads or common shoes with feet between the web of the shoe and the hoof, with sharp calks to take up the jar. It would be best not to shoe at all, but let the horse loose in a wet pasture for a good while.

CHAPTER XI

I N this chapter the author desires to give some hints about the treatment for diseases most common to horses.

COLIC

There are two kinds of colic, spasmodic and flatulent.

Spasmodic colic is known by the pains and cramps being spasmodic, in which there are moments of relief and the horse is quiet.

Flatulent colic is known by bloating symptoms and the pain is continual, the horse kicks, paws, tries to roll and lie on his back.

For spasmodic colic give ½ ounce laudanum, ½ pint whisky, ½ pint water; mix well and give in one does. If this does not help, repeat the dose in half an hour.

For flatulent colic give ½ ounce laudanum, ½ ounce turpentine, ½ pint raw linseed oil, ¼ ounce chloroform, ½ pint water. Mix well and give in one dose. Repeat in one hour if the pain is not relieved.

BOTS

Sometimes there is no other symptom than the bots seen in the dung, and in most cases no other treatment is needed than some purgative.

MANGE

Mange is a disease of the skin due to a class of insects that burrow in the skin, producing a terrible itch and scab, the hair falling off in patches, and the horse rubs against everything. After the affected parts have been washed in soap-water quite warm, dry and rub in the following: 4 ounces oil of tar, 6 ounces sulphur, 1 pint linseed oil.

LICE

Make a strong tea of tobacco and wash the horse with it.

WORMS

There are many kinds of worms. Three kinds of tape worms and seven kinds of other worms have been found in the horse. The tape worms are very seldom found in a horse and the other kinds are easily treated by the following: One dram of calomel, 1 dram of tartar emetic, 1 dram of sulphate of iron, 3 drams of linseed meal. Mix and give in one dose for a few days; then give a purgative. Repeat in three weeks to get rid of the young worms left in the bowels in the form of eggs, but which have since hatched out.

DISTEMPER

Distemper is a disease of the blood. The symptoms are: Swelling under the jaws; inability to swallow, a mucous discharge from the nose.

Give the horse a dry and warm place and nourishing food. Apply hot linseed poultice to the swellings under the jaws and give small doses of cleansing powder for a few days.

HYDROPHOBIA

As soon as a case is satisfactorily recognized, kill the horse, as there is no remedy yet discovered that will cure this terrible disease.

SPAVIN

There are four kinds of spavin and it is difficult for any one but a veterinarian to tell one kind from another. In all cases of spavin (except blood spavin) the horse will start lame, but after he gets warmed up the lameness disappears and he goes all right until stopped and cooled off, when he starts worse than before.

There are many so-called spavin cures on the market, some of them good, others worse than nothing. If you don't want to call a veterinarian, I would advise you to use "Kendall's Spavin Cure." This cure is one of the best ever gotten up for this disease, and no bad results will follow the use of it if it does not cure. It is for sale by most druggists.

In nearly all cases of lameness in the hind leg the seat of the disease will be found to be in the hock-joint, although many persons (not having had experience) locate the difficulty in the hip, simply because they cannot detect any swelling of the hock-joint; but

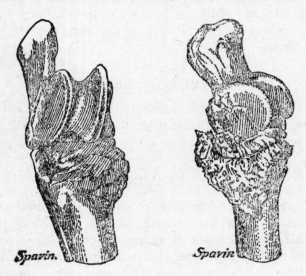

Spavin. *Spavin.*

in many of the worst cases there is not seen any swelling or enlargement for a long time, and perhaps never.

BONE SPAVIN

Bone spavin is a growth of irregular bony matter from the bones of the joint, and situated on the inside and in front of the joint.

Cause.—The causes of spavins are quite numerous,

but usually they are sprains, blows, hard work, and, in fact, any cause exciting inflammation of this part of the joint. Hereditary predisposition in horses is a frequent cause.

Symptoms.—The symptoms vary in different cases. In some horses the lameness comes on very gradually, while in others it comes on more rapidly. It is usually five to eight weeks before any enlargement appears. There is marked lameness when the horse starts out, but he usually gets over it after driving a short distance, and, if allowed to stand for awhile, will start lame again.

There is sometimes a reflected action, causing a little difference in the appearance over the hip joint, and if no enlargement has made its appearance, a person not having had experience is very liable to be deceived in regard to the true location of the difficulty. The horse will stand on either leg in resting in the stable, but when he is resting the lame leg he stands on the toe.

If the joint becomes consolidated the horse will be stiff in the leg, but may not have much pain.

Treatment.—That it may not be misunderstood in regard to what is meant by a cure, would say that to stop the lameness, and in most cases to remove the bunch on such cases as are not past any reasonable hopes of a cure.

But I do not mean to be understood that in a case of anchylosis (stiff joint), I can again restore the joint to its original condition; for this is an impossibility, owing to the union of the two bones, making them as one. Neither do I mean that, in any ordinary case of bone spavin which has become completely ossified (that is, the bunch become solid bone), that, in such a case, the enlargement will be removed.

In any bony growths, like spavin or ringbone, it will be exceedingly difficult to determine just when there is a sufficient deposit of phosphate of lime so that it is completely ossified, for the reason that in some cases

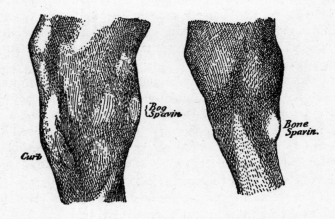

the lime is deposited faster than in others, and therefore one case may be completely ossified in a few months, while in another it will be as many years.

The cases which are not completely ossified are those that I claim to remove. One of this class which I have seen removed was a large bone spavin of four or five years standing, and I think that a large per cent of cases are not fully ossified for several months or years.

I am well aware that many good horsemen say that it is impossible to cure spavins, and, in fact, this has been the experience of horsemen until the discovery of Kendall's Spavin Cure. It is now known that the treatment which we recommend here will cure nearly every case of bone spavin which is not past any reasonable hopes of a cure, if the directions are followed, and the horse is properly used.

OCCULT SPAVIN

This is similar to bone spavin in its nature, the difference being that the location is within the joint, so that no enlargement is seen, which makes it more difficult to come to a definite conclusion as to its location, and consequently the horse is oftentimes blistered and tormented in nearly all parts of the leg but in the right place.

The causes and effects are the same as in bone spavin, and it should be treated in the same way.

These cases are often mistaken for hip disease, because no enlargement can be seen.

BOG SPAVIN

The location of this kind of a spavin is more in front of the hock-joint than that of bone spavin, and it is a soft and yet firm swelling. It does not generally cause lameness.

BLOOD SPAVIN

This is similar to bog spavin but more extended, and generally involves the front, inside and outside of the joint, giving it a rounded appearance. The swelling is soft and fluctuating. Young horses and colts, especially if driven or worked hard, are more liable to have this form of spavin than older horses.

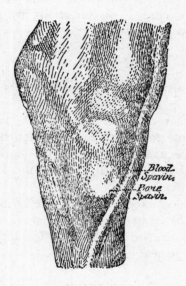

SPLINT

This is a small bony enlargement, and generally situated on the inside of the foreleg about three or four inches below the knee joint, and occurs frequently in young horses when they are worked too hard.

SPRAIN

By this is meant the sudden shifting of a joint farther than is natural, but not so as to produce dislocation.

Every joint is liable to sprain by the horse's falling, slipping, or being overworked. These cases cause a great deal of trouble, oftentimes producing lameness, pain, swelling, tenderness, and an unusual amount of heat in the part.

Treatment.—Entire rest should be given the horse, and if the part is found hot, as is usually the case, apply cold water cloths, changing frequently, for from one to three days until the heat has subsided, when apply Kendall's Spavin Cure, twice or three times a day, rubbing well with the hand.

If the fever is considerable, it might be well to give fifteen drops of tincture of aconite root, three times a day, for one or two days, while the cold water cloths are being applied. Allow the horse a rest of a few weeks, especially in bad cases, as it is very difficult to cure some of these cases, unless the horse is allowed to rest.

STAGGERS

A disease of horses, resulting from some lesion of the brain, which causes a loss of control of voluntary motion. As it generally occurs in fat horses which are well fed, those subject to these attacks should not be overfed. The cause is an undue amount of blood flowing to the brain.

Treatment.—The aim of the treatment should be to remove the cause. In ordinary cases give half a pound of epsom salts, and repeat if necessary to have it physic, and be careful about overfeeding.

In mad staggers, it would be well to bleed from the neck in addition to giving the epsom salts.

CERTAIN CURE FOR HOG CHOLERA

Take the following ingredients well mixed together, and give one tablespoonful daily in food during sickness, and as a preventative two or three times a week:

Powdered charcoal .... 1 pound
" mandrake .... 2 "
" resin .... 1 "
" saltpeter .... 8 ounces
" madder .... 8 "
" bi-carbonate of soda .... 6 pounds

TENSILE STRENGTH OF IRON AND OTHER MATERIALS

Pounds required to tear asunder a rod one inch square:

Cast steel 145,000
Soft steel 115,000
Swedish iron 85,000
American iron 60,000
Russian iron 62,000
Wrought wire 98,000
Cast iron, best..................... 45,000
Cast iron, poor..................... 14,000
Silver 40,000
Gold 21,000
Whalebone 8,000

Bone	8,000
Tin	5,000
Zinc	3,000
Platinum	40,000
Boiler plates	50,000
Leather belt (lin.)	350
Rope (manila)	10,000
Hemp (tarred)	14,000
Brass	40,000

HOW CORN IN THE CRIB AND HAY IN THE MOW SHOULD BE MEASURED

As near as can be figured out, two cubic feet of corn in the ear will make one bushel shelled. To find the quantity of corn in the crib, measure length, breadth and height, multiply the breadth by the length and this product by the height; then divide this product by two, and you have the right number of bushels of corn.

It is estimated that 510 cubic feet of hay in a mow will make one ton. Multiply the length by the breadth and the product by the height; divide this product by 510, and the quotient shows the tons of hay in the mow.

GRAIN SHRINKAGE

Not often do the farmers gain any by keeping the grain, for it will shrink more than the price will make good. Wheat will shrink 7 per cent in seven months from the time is is thrashed. Therefore, 93 cents a bushel for wheat in September is better than $1 in

April the following year. Add to this the interest for the money you could have used in paying debts, or loaned, and it will add 4 per cent more, making it 11 per cent.

Corn will shrink more than wheat, and potatoes are very risky to keep on account of the diseases they are subjected to; the loss is estimated at 30 per cent for six months.

VALUE OF A TON OF GOLD OR SILVER

A ton of gold is worth in money $602,799.23; a ton of silver, $37,704.84.

AGES OF ANIMALS

	Years.
Elephant	1 to 400
Whale	100
Swan	250
Eagle	100
Raven	110
Stag	50
Lion	75
Mule	75
Horse	30
Ox	30
Goose	75
Hawk	35
Crane	24
Skylark	20
Crocodile	100
Tortoise	150
Cow	20

Deer.....................................	20
Wolf.....................................	20
Swine....................................	20
Dog......................................	12
Hare.....................................	8
Squirrel..................................	7
Titlark...................................	5
Queen bee................................	4
Working bee......................6 months	

RINGWORM

Ringworm is a contagious disease and attacks all kinds of animals, but it often arises from poverty and filth. It first appears in a round bald spot, the scurf coming off in scales.

Cure: Wash with soap-water and dry, then apply the following once a day. Mix 25 grains of corrosive sublimate in half a pint of water and wash once a day till cured.

BALKING

Balking is the result of abuse. If a horse is overloaded and then whipped unmercifully to make the victim perform impossibilities, he will resent the abuse by balking.

There are many cruel methods for curing balking horses, but kindness is the best. Don't hitch him to a load he cannot easily pull. Let the man that is used to handling him drive him. Try to divert his mind from himself. Talk to him; pat him; give him a handful of oats or salt. But if there is no time to wait pass a chain or rope around his neck and pull him along with another horse. This done once all there is needed, in most cases, is to pass the rope around and the horse will start. It is no use trying to whip a balking horse, because balking horses are generally horses of more than common spirit and determination, and they will resent abuse every time. Kindness, patience and perseverance are the best remedies.

RATTLE-SNAKE BITE

When a horse has been bitten by a rattlesnake, copperhead, or other venomous serpent, give the following: One-half teaspoonful of hartshorn, 1 pint whisky, ½ pint of warm water. Mix well and give one dose. Repeat in one hour if not relieved. Burn the wound at once with a hot iron, and keep a sponge soaked in ammonia over the wound for a couple of hours.

HOOF OINTMENT

Rosin, 4 ounces; bees wax, 4 ounces; pine tar, 4 ounces; fish oil, 4 ounces; mutton tallow, 4 ounces. Mix and apply once a day.

PURGATIVE

Aloes, 3 drams; gamboge, 2 drams; ginger, 1 dram; gentian, 1 dram; molasses, enough to combine the above. Give in one dose, prepared in the form of a ball.

HINTS TO BLACKSMITHS AND HORSE-SHOERS

Don't burn the shoe on.
Don't rasp under the clinchers.
Don't rasp on the outer side of the wall more than is absolutely necessary.
Don't rasp or file the clinch heads.
Don't make the shoes too short. Don't make high calks. Don't pare the frog.
Don't cut down the bars. Don't load the horse down with iron.
Don't lose your temper. Don't hit the horse with the hammer.
Don't run down your competitor. Don't continually tell how smart you are.
Don't smoke while shoeing. Don't imbibe in the shop. Don't run outdoors while sweaty. Don't know it all. Always be punctual in attendance to your business. Allow your customers to know something. No man is such a great fool but that something can be learned of him.
Be always polite. Keep posted on everything belonging to your trade. Read much. Drink little. Take a bath once a week. Dress well. This done, the craft will be elevated, and the man respected.

ADVICE TO HORSE OWNERS

It is cruelty to animals to raise a colt and not train him for shoeing, and the horse-shoer must suffer for this neglect also. Many a valuable horse has been crippled or maltreated, and thousands of horse-shoers suffer hardships, and many are crippled, and a few killed every year for the horse owner's carelessness in this matter. A law should be enacted making the owner of an ill-bred horse responsible for the damage done to the horse-shoer by such an animal. Every horse-raiser should begin while the colt is only a few days old to drill him for the shoeing. The feet should be taken, one after the other, and held in the same position as a horse-shoer does, a light hammer or even the fist will do, to tap on the foot with, and the feet should be handled and manipulated in the same manner the horse-shoer does when shoeing. This practice should be kept up and repeated at least once a week and the colt when brought to the shop for shoeing will suffer no inconvenience. The horse-shoer's temper, as well as muscles, will be spared and a good feeling all around prevails.

Horse-raisers, remember this.

ADVICE TO YOUNG MEN

In every profession and trade it is a common thing to hear beginners say: I know, I know. No matter what you tell them, they will always answer, I know. Such an answer is never given by an old, learned or experienced man, because, as we grow older and wiser we know that there is no such thing as knowing it all. Besides this we know that there might be a better way than the way we have learned of doing the work. It is only in few cases that we can say that this is the best way, therefore we should never say, I know: first, because no young man ever had an experience wide enough to cover the whole thing; second, it is neither sensible nor polite. Better not say anything, but simply do what you have been told to do.

Every young man thinks, of course, that he has learned from the best men. This is selfish and foolish. You may have learned from the biggest botch in the country. Besides this, no matter how clever your master was, there will be things that somebody else has a better way of doing. I have heard an old good blacksmith say, that he had never had a helper but what he learned some good points from him.

Don't think it is a shame, or anything against you, to learn. We will all learn as long as we live, unless we are fools, because fools learn very little. Better to assume less than you know than to assume more.

Thousands of journeymen go idle because many a master would rather hire a greenhorn than hire a "knowing-it-all" fellow. Don't make yourself obnoxious by always telling how your boss used to do this or that. You may have learned it in the best way possible, but you may also have learned it in the most awkward way. First find our what your master wants, then do it, remembering there are sometimes many ways to accomplish the same thing. Don't be stubborn. Many mechanics are so stubborn that they will never change their ways of doing things, nor improve on either tools or ideas.

Don't be a one-idea man; and remember the maxim, "A wise man changes his mind, a fool never."

Be always punctual, have the same interest in doing good work and in drawing customers as you would were the business yours. Be always polite to the customers, no matter what happens. Never lose your temper or use profane language. Don't tell your master's competitors his way of doing business, or what is going on in his dealings with people. You are taking his money for your service, serve as you would be served.

IRON CEMENT

A cement for stopping clefts or fissure of iron vessels can be made of the following: Two ounces muriate of ammonia, 1 ounce of flowers of sulphur, and 1 pound of cast-iron filings or borings. Mix these well in a mortar, but keep the mortar dry. When the cement is wanted, take one part of this and twenty parts of clean iron borings, grind together in a mortar. Mix water to make a dough of proper consistence and apply between the cracks. This will be useful for flanges or joints of pipes and doors of steam engines.

HOW TO RUN A TURNING LATHE

(By a student of James College of Mechanic Arts, at Ames, Iowa.)

Lathes, when first invented, were very rude affairs, but they, like all other machinery, have experienced improvement from year to year until now some of them are more complicated than a watch, and for that reason should receive the best of care. They should be kept clean and well oiled. While being used the dust and shavings should be cleaned off at least every night, and every half day is better.

When they are kept in a dusty place, as is very often the case in a general repair shop, they should be kept covered while not in use. Some cheap canvas makes a good cover.

Every person who intends running a lathe should first become acquainted with his machine; become familiar with all the combinations that can be made, so that when a piece of work comes in to be done he will know just how to arrange the lathe to do that work. For instance, a piece of work needs to be turned tapering; this is done by shifting the tail stock to one side. Or there are threads to be cut; know just how to arrange the lathe to cut any number of threads to the inch.

Next to care of lathe comes care of tools. When there are a few minutes spare time see that the tools

are sharp. Keep them sharp. They will do the work better, faster and with much less strain on the machine.

All cutting tools should be made diamond shape, with either one side or the other, depending on the way the carrier is to move, made a little higher; the right side being highest when the carrier is moving to the right, and vice versa. The sharp edge of smoothing tools is made square across, like a plane bit, and thread-cutting tools should be made the same shape as the thread to be cut.

MODERN BLACKSMITHING

Water or oil should be kept on the iron or steel that is being turned. It keeps the point of the tool from getting hot when heavy chips are taken, and it makes a smoother job when the smoothing tool is used. There is no need to use either water or oil when turning cast iron.

The tempering of lathe tools is a very particular piece of work, varying considerable with the kind of steel used and the nature of the work to be done. For slow heavy turning the tool must not be too hard, else it would break; while for light swift turning it should be quite hard. For water tempering the temper color varies from a dark blue to a very light straw color, depending, as I have said before, on the nature of the work to be done.

By way of illustration of a piece of work that represents a number of lathe combinations, I will take the fitting of a saw shaft for our common wood saws. First place the balance wheel in the lathe chuck, being sure to get it in the center, so that when the hole is drilled in the wheel it will be in the exact center. Take a drill a sixteenth of an inch smaller than the hole to be made, and drill out the hole. Use the inside boring tool to make the hole the desired size. Turn a smooth face on the hub of the wheel where it comes against the box; then the wheel is ready for the key seat. To cut the key seat in the wheel use a key-seat chisel the same size as the milling wheel used to cut the key seat in the shaft.

Next take one of the saw collars; put it in the chuck, being careful to get this in the center also, with the widest side next the chuck, and drill a hole in it the same size as the hole in the saw. Turn off the end of the collar to get it square. Prepare the other collar in the same way.

Now cut the shaft off the length wanted, and turn one end to fit tightly into the balance wheel. Turn off a place next to where the wheel comes for the bearing or box. Now turn the shaft around and fit the other end for the collars. The collar that goes on the inside or side next the bearing should be shrunk on. To do this leave the shaft about one sixty-fourth of an inch larger than the hole in the collar, then heat the collar to a red heat, and slip it onto the shaft. It should not be driven very hard, or it will break in cooling. Let it cool of its own accord. When nearly cool it can be put into water and cooled off.

The next step is to true up the inside of the collar, leaving about one inch of surface to come against the saw. Now turn the shaft down to the size wanted for the thread, either 1-inch or 1⅛-inch, then with a cut-off tool about ⅛-inch wide, cut in next the shoulder the depth of the thread. If there is a die and tap handy that will be the quickest way to cut the thread, but if not handy then use the lathe. Now screw the nut on and turn off the inside of the nut. For fitting the loose collar there should be on hand a shaft about 14 or 16 inches long, turned a very little tapering; then drive the collar onto this shaft and finish it up. When ready put this collar into place on the saw shaft and screw the nut up tight. Now smooth off the outside of the collars for loops. Cut the key seat in the shaft and key the balance wheel on solid, being careful to get the distance between the wheel and the saw collar the exact distance between the outside of the boxes.

HOW TO BALANCE A PULLEY

When a pulley or balance wheel is to be balanced you must first have a shaft that is of the same size as the hole in the pulley. Of course, the wheel or pulley must be turned and trued up so that it is finished before you balance the same.

After the shaft has been put in and tightened, place two pieces of angle iron or T-iron about two feet long parallel on a pair of wooden horses. The irons must be level. Now place the pulley between the irons so that the shaft will have a chance to roll on the "T" or angle iron, and you will notice that the heaviest side of the pulley will be down. Start it rolling, and the pulley will always stop with the heaviest side down. Now, if the pulley or wheel, as the case may be, has a thick rim, then bore out from the heaviest side enough to balance, or you can drill a hole in the lightest side and bolt a piece of iron to it just heavy enough to balance the wheel.

HOW TO PUT IN A WOODEN AXLE

One of the most difficult pieces of work to do in a wagon shop is to put in a wooden axle.

In the first place, you must have well-seasoned timber, hickory or maple. Take out the old axle. The skeins will come off easy by heating them a little. Now cut the timber the exact length of the broken axle. In order to get the right pitch and gather, you must cut off one-half inch from the back side of the end of the timber and one-half inch from the bottom side, this cut to run out at the inner end or collar of the skein, as shown in Figure 14. Next take dividers and make a circle in the end of the axle the size of the old axle—in case new skein is put on, the size of the

bottom of the skein inside. This circle must be made so that the lower side of it will go down to edge of the timber, and the sides be of the same distance from the edges. You will now notice that most of the hewing will be done on top side, as it must in order to get the right pitch, and as one-half inch has been cut from the back side it will throw the front side of the wheel in a little; this is gather. If a wheel has no gather the wheel will be spread out against the nut of the skein, and the wear will be in that direction, and the wheel will rattle, as you know the skein is tapered; but if the wheel has gather, the pressure will be against the collar of the skein, and the wheel will be tight, as it forces itself up against the collar and the wider end of the skein.

Some wagon-makers will use the old axle as a guide and cut the new by the old. This is not safe, as the old is mostly sprung out of shape.

In hewing the axle for the skein great care should be taken not to cut off too much; better go slow, because it depends upon the fitting of the skein to get a good job. When the axle is finished or ready to be driven into the skein be sure to have the axle strong; that is, a little too large to go in easy. Now warm—or heat, if you will—the skein a little, not so much that it will burn, and drive it onto its place by a mallet. In making new wagons I think it would be wise to paint the part of the axle that goes in the skein, but in repairing I deem it unwise, because it will have a tendency to work loose unless it will have time to dry before using, and I have noticed paint to be still fresh in the skein after years of use. There should be no gap left between the collar of the skein and the axle, as water will run in and rot the timber.

HOW TO PUT IN SPOKES

VERY wagon-maker is supposed to know how to put in spokes. Still, there are sometimes wagon-makers, especially beginners, that don't know. First clean out the sliver left of the old spoke, and make the mortise dry, and in every case use glue. In a buggy wheel take the rivet or rivets out, if there is any, and be sure to have the right shape of the tenon to fit the mortise in the hub, so as to make the spoke stand plumb. Set the tenon going through the rim. Be sure to have this tenon reach through. This is important in filling a wagon wheel, because, if the tenons don't reach through the fellow, then the heft will rest against the shoulder of the tenon, and when the tire is put on tight and the wagon used in wet roads, the fellow will soften and the spokes settle into the rim. The tire gets loose, and some one, either the wagon-maker or the blacksmith, will be blamed—in most cases the blacksmith. Of course, the tenon should not be above the rim. After the spokes have been put in rivet the flange of the hub, or so many rivets as you have taken out. This should always be done before the tire is set.

WEIGHT OF ONE FOOT IN LENGTH OF SQUARE AND ROUND BAR IRON

Size.	Square.	Round.	Size.	Square.	Round.
1/4	.209	.164	2 1/8	15.000	11.840
5/16	.326	.256	2 1/4	16.900	13.280
3/8	.469	.368	2 3/8	18.835	14.792
7/16	.638	.504	2 1/2	20.871	16.392
1/2	.833	.654	2 5/8	23.112	18.142
9/16	1.057	.831	2 3/4	25.250	19.840
5/8	1.305	1.025	2 7/8	27.600	21.681
11/16	1.579	1.241	3	30.065	23.650
3/4	1.875	1.473	3 1/8	32.610	25.615
13/16	2.201	1.728	3 1/4	35.270	27.702
7/8	2.552	2.004	3 3/8	38.040	29.875
15/16	2.930	2.301	3 1/2	40.900	32.160
1	3.340	2.625	3 5/8	43.860	34.470
1 1/8	4.222	3.320	3 3/4	46.960	36.890
1 1/4	5.215	4.098	3 7/8	50.150	39.390
1 3/8	6.310	4.960	4	53.435	41.980
1 1/2	7.508	5.900	4 1/4	60.320	47.380
1 5/8	8.810	6.920	4 1/2	67.635	53.130
1 3/4	10.200	8.040	4 3/4	75.350	59.185
1 7/8	11.740	9.222	5	83.505	65.585
2	13.300	10.490	6	120.240	94.608

WEIGHTS OF ONE LINEAL FOOT OF FLAT BAR IRON

Thickness.	Width, 1.	Width, 1¼.	Width, 1½.	Width, 1¾.
⅛	.416	.521	.624	.728
3/16	.625	.780	.938	1.090
¼	.833	1.040	1.250	1.461
5/16	1.041	1.301	1.560	1.821
⅜	1.252	1.562	1.881	2.190
7/16	1.462	1.822	2.191	2.550
½	1.675	2.085	2.505	2.925
9/16	1.884	2.345	2.815	3.285
⅝	2.085	2.605	3.132	3.655
11/16	2.295	2.860	3.442	4.010
¾	2.502	3.131	3.752	4.381
⅞	2.921	3.650	4.382	5.100
1	3.331	4.170	5.005	5.832
1⅛	3.750	4.694	5.630	6.560
1¼	4.175	5.210	6.251	7.290
1⅜	4.580	5.728	6.879	8.022
1½	5.005	6.248	7.502	8.750
1⅝	5.425	6.769	8.130	9.480
1¾	5.832	7.289	8.749	10.208
1⅞	6.248	7.800	9.380	10.938
2	6.675	8.332	10.005	11.675

WEIGHTS OF ONE LINEAL FOOT OF FLAT BAR IRON

(Continued)

Thickness.	Width, 2.	Width, 2¼.	Width, 2½.	Width, 2¾.
⅛	.832	9.370	1.040	1.151
3/16	1.251	1.410	1.562	1.720
¼	1.675	1.878	2.080	2.290
5/16	2.081	2.342	2.000	2.862
⅜	2.502	2.811	3.135	3.445
7/16	2.920	3.278	3.650	4.010
½	3.335	3.748	4.175	4.580
9/16	3.748	4.220	4.089	5.160
⅝	4.168	4.690	5.211	5.730
11/16	4.578	5.160	5.735	6.150
¾	5.005	5.630	6.255	6.880
⅞	5.830	6.558	7.395	8.025
1	6.668	7.500	8.332	9.170
1⅛	7.498	8.441	9.382	10.310
1¼	8.333	9.382	10.421	11.460
1⅜	9.775	10.310	11.460	12.605
1½	10.000	11.255	12.505	13.750
1⅝	10.835	12.190	13.545	14.905
1¾	11.675	13.135	14.585	16.045
1⅞	12.505	14.065	15.635	17.195
2	13.335	15.000	16.675	18.335

WEIGHTS OF ONE LINEAL FOOT OF FLAT BAR IRON

(Continued)

Thickness.	Width, 3.	Width, 3¼.	Width, 3½.	Width, 3¾.
⅛	1.250	1.350	1.465	1.658
3/16	1.879	2.035	2.195	2.345
¼	2.505	2.710	2.925	3.135
5/16	3.135	3.391	3.650	3.901
⅜	3.750	4.060	4.380	4.695
7/16	4.385	4.740	5.105	5.470
½	5.000	5.425	5.832	6.250
9/16	5.635	6.090	6.565	7.030
⅝	6.255	6.775	7.290	7.805
11/16	6.885	7.455	8.020	8.590
¾	7.500	8.135	8.750	9.380
⅞	8.750	9.480	10.210	10.940
1	10.000	10.835	11.675	12.500
1⅛	11.255	12.190	13.135	14.065
1¼	12.505	13.540	14.585	15.635
1⅜	13.750	14.905	16.045	17.195
1½	15.000	16.250	17.500	18.750
1⅝	16.255	17.605	18.960	20.310
1¾	17.505	18.965	20.425	21.880
1⅞	18.750	20.305	21.885	23.445
2	20.000	21.670	23.335	25.000

WEIGHTS OF ONE LINEAL FOOT OF FLAT BAR IRON

(Continued)

Thickness.	Width, 4.	Width, 4¼.	Width, 4½.	Width, 4¾.
⅛	1.670	1.774	1.887	1.989
3/16	2.500	2.658	2.811	2.971
¼	3.331	3.538	3.750	3.960
5/16	4.168	4.430	4.689	4.950
⅜	5.000	5.311	5.630	5.940
7/16	5.831	6.200	6.560	6.930
½	6.670	7.082	7.502	7.925
9/16	7.500	7.965	8.435	8.910
⅝	8.330	8.855	9.380	9.900
11/16	9.165	9.740	10.310	10.890
¾	10.000	10.630	11.250	11.880
⅞	11.670	12.400	13.140	13.845
1	13.340	14.165	15.000	15.830
1⅛	15.000	15.940	16.880	17.815
1¼	16.660	17.710	18.755	19.179
1⅜	18.335	19.480	20.650	21.770
1½	20.000	21.255	22.505	23.750
⅝	21.675	23.025	24.380	25.730
1¾	23.335	24.790	26.240	27.710
1⅞	25.000	26.560	28.140	29.000
2	26.670	28.335	30.000	31.670

INDEX.

	PAGE
Advice to Horse Owners	180
Advice to Young Men	181
Anatomy of the Horse	135
Annealing	116
Anvil, The	33
Axle Gauge	85
Axes and Hatchets	67
Babbitting	114
Bands or Hoops	57
Back Dished Wheel	75
Belts, Points on	65
Blacksmith's Tongs	39
Blowers	54
Blowing out the Boiler	131
Bob Shoes	66
Case Hardening	123
Coal Box, The	34
Cold Chisels	43
Drilling Iron	58
Expansion of the Tire	83

INDEX.

	PAGE.
Fire, The	50
Forge, The	33
Forging	143
Foaming in Boilers	130
Gather Gauge	85
Grain Shrinkage	175
Hammer, The	35
Hints to Blacksmiths	179
Hints to Horseshoers	179
Horse, The	132
Horseshoeing	134
How to Measure Corn in the Crib and Hay in the Mow	175
How to Make a Landside	88
How to Harden Springs	124
How to Weld Cast Iron	121
How to Repair Broken Iron Pump Handles	123
How to Repair Broken Cogs	116
How to Restore Overheated Steel	117
How to Dress and Harden Stone Hammers	117
How to Drill Chilled Cast Iron	118
How to Drill Hard Steel	118
How to Make Steel and Iron as White as Silver	124
How to Mend Broken Saws	125
How to Mend a Band Saw	125
How to Write Your Name on Steel	126
How to Patch a Boiler	127
How to Put in Flues	128
How to Weld Flues	129
How to Make the Shoe	139
How to Prepare the Foot for the Shoe	142

INDEX.

	PAGE.
How to Shoe a Kicking Horse	160
How to Shoe a Trotter	162
How to Shoe a Horse with Brittle Hoofs	163
How to Shoe a Weak Heeled Horse	163
How to Shoe a Knee Sprung or Knuckler	149
How to Run a Lathe	183
How to Balance a Pulley	187
How to Put in a Wooden Axle	187
How to Put in Spokes	190
How to Strike and Turn the Iron	49
How to Make a Hammer	41
How to Make Chisels	43
How to Harden Files	62
How to Harden Taps and Dies	62
How to Make Butcher Knives	63
How to Repair Cracked Circular Saws	63
How to Prevent a Circular Saw from Cracking	64
How to Sew a Belt	64
How to Drill Chilled Iron	59
How to Make Plowshares	88
How to Put on New Tire	76
How to Weld Tires	77
How to Harden a Plowshare	103
How to Point a Share	104
How to Sharpen a Plowshare	107
How to Put on a Heel	108
How to Repair a Flopping Plow	108
How to Set a Plow Right	109
How to Correct Plow from Running too Deep	109
How to Fix a Gang Plow that Runs on Its Nose	110

INDEX.

	PAGE.
How to Harden a Mouldboard	111
How to Patch a Mouldboard	111
Incompetency	17
Interfering	146
Intemperance	14
Iron Cement	182
Landside Point for Slipshare	92
Literature	27
Mill Picks	61
Modern Guild	19
Mower Sections	114
Religion	16
Rock Drills	47
Rules for Smith and Helper	49
Setting Tire	72
Seeder Shovels	57
Set Hammer	44
Shoe, Right Fitting	144
Shoe, Wrong Fitting	144
Shop, The	31
Sledge, The	36
Slipshare	99
Smith, The	9
Split Welds	56
Standing Coulters	59
Steel, Facts about	119
S Wrench	47
Taxation	21
Tensile Strength of Iron and other Materials	174

INDEX.

	PAGE.
Tire in Sections	82
Toe Tips	144
Tools, Granite	69
Tool Table	34
Tuyer Iron	52
Twist Drills	45
Vehicles	71
Wagon Making	71
Water Tuyer	53
Welding Axles	84
Welding Steel	56
Welding Iron	55
Weight of One Foot in Length of Square and Round Bar Iron	192
Well Drills	68
Diseases of the Horse	165
Bots	166
Mange	166
Lice	166
Worms	166
Distemper	167
Hydrophobia	167
Spavin	167
Bone Spavin	168
Occult Spavin	171
Ages of Animals	176
Ring Worms	177
Balking	177
Founder	164

INDEX.

	PAGE.
Hoof Ointment	178
Purgative	178
Horse, The Wall	137
Horse, The Sole	138
Horse, The Frog	138
Horse, The Coronet	139
Horse, The Quarter	139
Horse, The Bars	139
Contraction	150
Corns	155
Quarter and Sand Cracks	157
Seedy Toe	158
Pricking	158
Stifled	159
String Halt	159
Bog Spavin	171
Blood Spavin	171
Splint	172
Sprain	172
Staggers	173
Hog Cholera, Cure for	174

TWENTIETH CENTURY
TOOLSMITH
AND
STEELWORKER

CONTENTS.

Introduction ....Pages 11 to 14

CHAPTER I.

Steel, its use and necessity in all arts, trades and professions—The composition of cast tool steel—The successful treatment of steel—Heating—Forging—Hammering—Hardening—The hardening bath—Quenching—Tempering — Welding — Annealing — Different kinds of steel—The cracking of tools when hardening and the cause—How to judge hard from soft steel—How to tell good from poor steel—Testing steel after hardening, with a file—Instructions on toolmaking that have to be given many times........
Pages 15 to 31

CHAPTER II.

The blacksmith's fire—Bellows and blowers—The anvil—Tongs—Fullers and swages—Flatters and set hammer—The hammer—Making and dressing a hand hammer—Hardening and tempering a hammer—Successful points to be remembered in making and tempering a hammer—Punching holes in steel. Pages 32 to 55

CHAPTER III.

The cold chisel—The hardy—Heavy hot, cold and railroad chisels—Drills and drilling—Making a flat drill—Hand made twist drills—Making a twist reamer..
Pages 56 to 72

CONTENTS

CHAPTER IV.

How to draw out, harden and temper an axe that will cleave a hemlock knot—Mill picks—Butcher knives—How to make gun, revolver, trap and all fine springs—Dirt picks—Laying dirt picks.........Pages 73 to 89

CHAPTER V.

Machinists tools—Air hardening steel for lathe and planer tools—How to anneal air hardening steel—Milling cutters—The use of asbestos and clay when hardening milling cutters and other tools—Hardening hollow tools—The hardening and tempering of hob taps, stay bolt taps ana similar tools—Heating furnaces—Heated lead for hardening purposes—Boilermakers' tools—The beading tool—Punches and dies—Flue expanders—Drifts, rivet snaps, calking tools and chisels—Hardening shear blades....Pages 90 to 115

CHAPTER VI.

Woodworkers' and carpenters' tools—Laying a carpenter's chisel—The screw driver—How to make a draw knife ....Pages 116 to 121

CHAPTER VII.

Granite cutters' tools—The granite drill—Bull sets and bull chisels—The granite bush hammer—The granite hammer—Granite cutters' mash hammer—The granite tool sharpeners' hammer and anvil stake—Marble cutters' tools ....Pages 122 to 130

CONTENTS

CHAPTER VIII.

Limestone cutters' tools—Plain and tooth chisels and points—Pitching tool—Hand and ball drills—The tooth axe—The limestone bush hammer—Sandstone cutters' tools—Stone carvers' tools—Polishing board for stonecutters' tools—How to forge mallet head tools—Punching teeth in stone cutters' tools—Lathe and planer tools for cutting soft stone—Dressing tools with the cutting edge bevelled from one side only ....Pages 131 to 145

CHAPTER IX.

The stonemason's hammer—Miners' tools—Correct and incorrect shapes of miners' hand drills—The cross or machine drill—The breaking of drills when drilling and the cause—The rock cutting reamer—Well drills
Pages 146 to 153

CHAPTER X.

Horseshoer's tools—How to make and dress a pair of pincers—Making a clinch cutter—How to make a horseshoer's knife—How to dress a vise—Sharpening plow shares—How to make square holes in plow shares ....Page 154 to 162

CHAPTER XI.

How to make a harnessmaker's knife—How to make a butcher's steel—Hardening tools with five projections—The butcher's cleaver—How to dress a railroad pinch bar—The spike maul—The claw bar........
Pages 163 to 173

CONTENTS

CHAPTER XII.

The bricklayer's set—How to harden and temper wire nippers or pliers—How to make a razor—To make a scraper—Hardening jaw of pipe vise—Hardening and tempering blacksmith's bolt clipper—Tools for punching or gumming cross-cut saws—The scratch-awl—Hardening and tempering circular blades of pipe cutter—Hardening a tool according to its shape—Making, hardening and tempering an alligator pipe wrench—Hardening and tempering pruning shear blades—The center punch—The nail set—Hardening and tempering steel stamps—Making a gouge—Hardening and tempering carpenters' augurs that have come through a fire—Case hardening..........
Pages 174 to 187

CHAPTER XIII.

The correct meaning of a cherry red heat—Heating to harden according to the size of the tool—Charcoal for heating steel—The sealing of steel after hardening—Quality and quantity—Quick methods of working—Cracks in steel—Slighting tools—The result of being a successful steelworker—Hardening tools that are forged by another mechanic—Sayings and ideas of mechanics in reference to steel—Why some tools are soft when put into use—Reasons why tools break when in use—Necessary tools—Welding compounds—Hardening compounds—How to determine the temper of tools—Overheating tools—Cutting steel when cold—Oil tempering—Drawing the temper over the open or blacksmith's fire—More points on hammering steel—How to improve—The blacksmith's helper—The danger of heating more of a tool when dressing, than what is to be forged or hammered—Hardening very small or thin tools—More information about

CONTENTS

cold chisels—The different degrees of temperature Fahrenheit required to equal the various colors when drawing the temper in hot air or oil—Table of ordinary tools made from cast steel, arranged alphabetically, giving the color of temper and about the percentage of carbon the steel should contain to give the best results—Table of tools continued, which are partly or wholly hardened but have no temper drawn—Working steel at night—A few words in reference to burnt steel—Conclusion ....Pages 188 to 229

Useful Formulas ....Pages 230 to 240

INTRODUCTION

In introducing this book to my readers and brother mechanics, it has long been my aim to bring this volume of information before the steelworkers and toolmakers in general and to present it in a clear simple way that the average mechanic will quite readily understand. Although there are other books written and published on this great subject of steelworking, the information which they contain is not expressed in a clear light that is beneficial to the ordinary reader, for unless the reader is already an expert steelworker the book is not easily understood without a great deal of thought and study, as some authors oppose their own ideas, others again do not take up the entire subject and the information which is most profitable to the young mechanic and also which is most impressive on his mind is left unwritten.

But this book which is entitled, The 20th Century Toolsmith and Steelworker, will give fully all the information and knowledge of working steel in a clear light so that the young mechanic or apprentice will readily improve, if he will but read. The methods given and used as regards the working of steel, are of the most modern, simple, practical, and scientific nature, while the instructions are from the experience of a successful steelworker of good reputation, and who has spent years in hard work, ranging in extent from the humble country blacksmith shop to the largest and best railroad, locomotive and machine shops, also stone yards, quarries and mines of North America, which is the only correct way of gathering together the vast amount of knowledge contained in these pages and which has cost the author thousands of dollars in wide travel and collecting valuable ideas from some of the greatest living mechanics and steelworkers that America has produced.

Although this book is chiefly intended for blacksmiths, toolsmiths and tooldressers, it will be found invaluable to every mechanic connected directly or indirectly with the repair and manufacture of steel into tools, and if the directions are followed closely, the amateur steelworker will become an expert of the highest degree, as there is nothing mentioned, but that which has been accomplished by the author and proven by experiment to be the greatest success.

This book is not merely written for the young mechanic or apprentice, but likewise for the old, and it does not signify if the reader has worked over the anvil for forty years, there is information that will help him overcome difficulties and obstacles connected with steel.

Although the instructions given are principally in reference to heating steel in the blacksmiths or open fire and which is mostly used, this book gives informa-

INTRODUCTION 13

tion concerning heating and tempering furnaces. But, it should be remembered that if a mechanic can work steel by heating it successfully in the open fire, he will experience very little trouble when heating steel in a furnace or lead bath as used in large and up-to-date toolshops and factories.

I wish to say to all mechanics young and old but more so to the young mechanics who have a desire to reach the top of the ladder and gain a good reputation, and especially to those who chance to get a copy of this book, that the greatest obstacle they have to contend with when trying to improve, is to change from the rut they have already fallen into, chiefly made by themselves and the teachings of their first masters. I state this from experience, and to illustrate fully after I knew my trade (or "served my time," is a more reasonable way of explaining), having plenty of confidence and a great share of conceit in my abilities, I started out as a journeyman blacksmith, "and then" I found out I had something to learn. But I found out that to change my ways and ideas was quite a difficult task and often got me in hot water, as I thought my way or rather the way I was taught by my first boss was correct. However, I soon decided that if I wanted to climb to the top of mechanical success and have a good reputation, I would have to change my ideas if I thought some other shop mate had an idea or method that was superior to mine, keep my eyes open, and do a great deal of thinking in my "own" mind.

And if I could have had this book at the close of my apprenticeship, it would have saved me many a troublesome job, many a long hour of study, a great deal of experimenting, large sums of money and placed me years in advance of the present times. And so I wish to say to the reader, although he may have some good ideas that perhaps are equal to the author's, while on the other hand he may have some not as good, read this book carefully from beginning to end, and follow its advice and he will be crowned with success, as a poor mechanic or Jack of all trades is not wanted in these days, where there is as much competition for the mechanic's job as there is between business men in any mercantile business. And again, I say to the mechanic read this book carefully, follow the instructions closely, and you will hold your job and take first place.

CHAPTER I.

Steel, its use and necessity—Composition—Successful treatment—Different kinds—The cracking of steel when hardening and the cause—Judging and testing.

Steel, Its Use and Necessity in All Arts, Trades and Professions.

We could get along without a great many other materials and metals, "but we must have steel," its great necessity comes first in all arts, trades, and professions. The doctor or surgeon must have fine lancing knives, the dentist must have forceps, and the sculptor must have fine chisels. The machinist, boilermaker, stonecutter, bricklayer and the stone mason, must have their tools made from steel in order to perform their skillful labor, likewise, the king and foundation of all mechanics, "The Blacksmith," he must in the first place have the anvil, hammers, chisels, fullers, swages, etc., to do his own work and make tools for others.

Our capitols, government buildings, palaces, cathedrals, the great railroad systems, likewise the defense of our country, the navy and its guns, are all brought to the stage of perfection, by the use of steel and so we can readily see that steel comes first and foremost of all metals, and the mechanic who is so fortunate as to become a good steelworker, is entitled to all honor and should be proud of his skill.

The Composition of Cast Tool Steel.

In order to understand this subject fully, it is necessary to know something pertaining to the manufacture. But in a simple way of explaining, cast tool steel is chiefly composed of bar or wrought iron, although wrought iron is a very useful metal, it is of too soft a nature in its natural state for the purpose of tool making. Consequently iron is put through a process by the steel manufacturer, and by the use of charcoal the iron becomes carbonized and so converted into steel.

A great deal could be written on the manufacture of cast tool steel, as the steel, after being manufactured may be of good or bad quality and also being of different degrees of hardness or temper. The quality of the steel depends on the quality of iron used in the manufacture, while the hardness of the steel depends on the amount of carbon it contains. The temper of steel is classed or measured by the percentage of carbon in the steel, for example 100 points is equal to 1 per cent, to further explain, steel that is right for making cold chisels will consist of 75 points carbon, while steel used for machinists' lathe or turning tools which is required to be much harder will consist of 1 per cent.

Good cast steel should be manufactured from pure Swedish iron and should contain not less than 60 points carbon in order that it will readily become (after passing through the various processes of the steelworker) hard as glass, tough as whalebone and as soft as lead. When toolmaking, a good quality of steel should always be used, but as to hardness, it will depend on how the tool is to be used and what material the tool is to cut or be used upon, also a great deal will depend on the skill of the steelworker.

The Successful Treatment of Steel.

In the successful treatment of steel lies the foundation of this book and the toolsmith's art, and with which all the following processes that the steel must pass through before reaching the finished tool are connected. Thus: heating, forging, hammering, hardening, quenching, tempering, welding and annealing. These processes all form an equally important part in the manufacture of tools, and so to become an expert steelworker this subject must be understood by having a thorough knowledge regarding the nature of steel, together with good judgment, carefulness and skill.

Heating.

There are a great many different ways of heating steel, although the most common way is in the blacksmith's coal or open fire, but in the large shops where tools are made in great quantities may be found furnaces especially adapted for tool making which are heated by oil, gas, etc.

The heating of steel is somewhat complicated owing to the different temperatures that are required (reader, give this your particular attention as success depends on the following), heating may be divided into four classes, as the forging heat, the hardening heat, the annealing heat and the welding heat. But for the benefit of the apprentice, I will say the different heats must be learnt by experimenting, but to the blacksmith of more or less experience, I will describe the heats of a piece of ordinary steel 75 points in carbon (which will answer all ordinary purposes), in the following manner, a yellow heat for forging, cherry red for hardening, blood red for annealing, and a white heat for welding.

Forging.

Forging is the toolsmith's labor which is required to bring or change a piece of steel into any shape or form, by referring to the forging heat, it is at all times necessary and beneficial to have a yellow or soft heat, then the steel will be worked clear through, and especially in heavy forging, but the heat must gradually decrease as the tool becomes finished. For example, supposing the toolsmith has a piece of steel one inch square and it is to be forged down to a chisel shaped point, it is heated slowly and evenly to a high or yellow heat, the toolsmith and helper forges it into the shape required until it is necessary to get another heat, but the second heat will not necessarily be so hot as it will be sufficient to finish the tool and the hardest work is over, when the steel is finished at a low heat and the last blows of the hammer fall on the flat side. The steel is left finer and stronger than if finished at a high heat.

Hammering.

Hammering steel in the finishing stage is one of the greatest secrets of success connected with forging tools, it is at all times necessary as it toughens, refines and packs the steel, but it is chiefly for tools that have a flat surface. On tools that have no flat surface but are either round or square, the blows must naturally fall on all sides alike, consequently the steel is left in its natural state. But tools that are flat, such as cold chisels or mill picks, the last blows say 10 or 15, must fall on both flat sides evenly when at a low heat, but bear in mind that not a blow is to strike the edge as it will knock out all the tenacity that has been put in the steel by the blows on the flat surface, and do not hammer the steel too cold as it will ruin the steel. If the hammering is properly done the steel will show a bright black gloss.

Hardening.

The process for hardening is by heating the steel to a certain heat then cooling off suddenly in water, which will immediately change the steel from its soft natural state, into that of a hard glasslike state and will show a white appearance when coming out of the water. But after coming through this operation the steel may be properly or improperly hardened, steel that is properly hardened is finer and stronger than improperly hardened steel, and if broken would present a fine crystalized fracture, while on the other hand, improperly hardened steel when broken would present a coarse fracture resembling a piece of honeycomb and will break very easy. The secret of success for proper hardening lies in the heat that is used, the proper heat must be found out by experimenting. A good way to find out the proper heat, will be, take a small piece of steel and on one end put deep nicks in it with a chisel, about half an inch apart, say, for three inches back, as shown in Figure 1. Now place the end that has the nicks, in the fire and heat the extreme point to a white or welding heat, then plunge into cold water and cool off "dead cold." Now place over the anvil, commencing at the extreme point that was the hottest and break off at the first nick, then the next and so on until all is broken, and the results will be as formerly explained. The first piece when broken will show a coarse, hard and very brittle fracture being very easy to break, and as the other pieces are broken the fracture will be noticed to be getting finer and harder to break until the

Fig. 1. Showing piece of steel for hardening test.

one is come to showing a fine crystallized fracture resembling a piece of glass. Another way to find out the proper heat for hardening will be to have 3 or 4 thin flat pieces of steel, heat them to different heats then break off taking particular notice of the fracture, and how some pieces will break much easier than others.

When hardening steel, always bear in mind to harden it at as low a heat as it will be sure to harden at, as proper hardening is the gateway of success in making tools that have to hold a good cutting edge.

The Hardening Bath.

In connection with the hardening of steel the hardening bath forms a very important part and which should not be overlooked. It consists chiefly of water, which must be clean and free from all oily or greasy substance. Water containing any greasy substance of any kind will not act so quickly or as satisfactory as clear, clean water. Rain or soft water is preferred to hard well water, but salt put in the water to form a brine is still better, as steel will harden at a lower heat in brine than in the ordinary pure water and this is a point to be well considered, so keep as much salt in the water as it will dissolve or soak up. Still another advantage by the use of brine is that it will not steam up so quickly as water and this is also worthy of thought when hardening large tools. At all times keep the bath as large and as cold as possible.

Quenching.

In the quenching or cooling of steel during the hardening process, a great deal is to be learned, as sometimes the tool is liable to warp when being quenched, in some cases so bad as to spoil or crack the tool, while the cause will occur from improper quenching, as a great deal depends according to the way the tool is placed in the water or hardening bath, and also according to the shape of the tool.

Some tools must cut the water as a knife, others again must thrust it as a dagger, and some at one angle, some at another. For example, take a round piece of steel 6 inches long and $\frac{3}{4}$ thick, and it is to be hardened the whole length of itself. After heating, it will have to be lowered into the bath from a perfectly upright position, if it has been properly forged, heated, and annealed, it will come from the water perfectly straight, but should it be placed in the water from an angular position it will be very apt to warp. Wide flat tools, whether partly or wholly hardened should be quenched in a perfectly upright position or they will warp flatwise.

Tempering.

After hardening the steel it will be too hard for some purposes, and so the hardness must be reduced by reheating it to a certain degree according to the work it is to do, which is "termed" tempering. If a piece of hardened steel be polished bright, then reheated, different colors will appear and change as the steel becomes heated to a greater degree. The colors will appear in their turn as follows, commencing with the least degree of heat will be a light straw, dark straw, copper, red, purple, dark blue, light blue and grey, and by watching the colors the steelworker regulates the temper or hardness of the tool.

Tempering is the process that will readily change steel from its hardened glass like state into an elastic springy nature resembling whalebone. For illustration, take a thin piece of steel 3 inches long, ½ inch wide, 1-16 thick, after hardening the whole piece from end to end then tempering to a very light blue and allowing it to cool off on its own accord, it will be found to be in a very elastic state and if bent it would immediately come back straight again. Tempering should not be classed as hardening or vice versa, as is often the case with a great many mechanics. For example, a tool that is to be only hardened and no temper drawn, should be classed as hardened, "and not tempered."

Welding.

Welding is the process or art of joining two pieces of steel together so as to form one solid piece, and which forms a very important part in steel working or toolmaking. There are several errors made when welding steel, some of the most common ones being, the want of the proper knowledge concerning the nature of steel, a green or unclean fire containing sulphur and other foreign matter, which is dangerous to hot steel, the absence of the proper welding heat, and improper ways of uniting the pieces together. For the benefit of those who have not had much practice and those who have only been partly successful, I will give these instructions, which, if followed closely will insure success. First of all the welding point in the steel that is to be welded must be known, as there are several kinds of steel, some will require a higher heat to weld than others, the heat varying according to the hardness of the steel.

For illustration, we make a weld by uniting two pieces of steel together and we have had good success, as the weld represents one solid piece. Now we proceed to make another weld, and in exactly the same way as the first weld, the same welding heat is used and the same fire, but this time we do not meet with success for as soon as the hot steel is struck with the hammer, to form the weld, the steel flies to pieces (I hear the reader ask the reason, why), because the steel was heated to a higher heat than what the steel would stand, and the consequence is, all the labor has been lost, the fault lies in not knowing the welding point.

We could take wrought iron and make every weld at

the same heat, but not so with steel, on account of it varying in hardness. And so in cases when the mechanic is in doubt as to the hardness or welding point in the steel, use this rule. Take a piece of the steel that is to be welded, heat it to a yellow heat, then place it over the edge of the anvil and strike it a light blow with the hammer, if the steel does not crumble or fall to pieces, keep increasing the heat until it does, this will enable anyone to test the steel for hardness, and so find the welding point or just how high a heat the steel will stand before crumbling or flying to pieces when making a weld.

But although the welding heat is well understood, there are other things to consider, as we must have a clean fire with the coal well charred and all gas, sulphur, clinkers, ashes, etc., must be taken from the fire, to insure a solid weld. Welding is more fully explained in another chapter of the book, as, in dirt picks.

Annealing.

The chief object of annealing steel is to soften it, the process being almost opposite to that of hardening. In hardening, the steel is cooled off very quickly, but in annealing, the steel is cooled very slowly. Steel to anneal must be heated in somewhat the same manner as to harden, with the exception that the annealing heat must not exceed the proper hardening heat, a little less heat will be best, for example a blood red.

The advantage to be gained by annealing steel is to make it soft, in order that it may be easily filed, turned, or planed. Without annealing some steel will be too hard for the machinist's use, tools that are forged by the toolsmith and finished by a machinist should always be annealed, and in a great many cases the steel must be annealed when it comes from the manufacturer, before it can be worked satisfactorily.

There are many ways to anneal, but the method that is commonly used, is by taking a piece of steel heated to the heat previously mentioned, and packed deep into slack lime allowing it to remain there until perfectly cold. Wood ashes may be used in place of slack lime, but they should be perfectly dry and free from all dampness. Fine dry sawdust is also very good, but it should be kept in an iron box in case the sawdust catches fire.

There is another good way to anneal and which is very often preferred on account that it is much quicker. Take a piece of steel heated as mentioned. Then hold it in a dark place long enough, so that the heat will all pass off, save a dim dull red. Then plunge into water to cool off. This is called the water anneal, and some machinists say that tools take a better hold of it. If the process was right the steel will come from the water resembling a piece of hardened steel, showing a black and white appearance by being partly scaled off. This method, however, may need a little experimenting before getting the best results. Points on annealing will be found in other parts of the book.

Different Kinds of Steel.

There are many kinds, grades, and brands of steel which vary in shape, quality, and hardness, according to the tool that is to be made from it and which the ordinary blacksmith is not familiar with. Steel used in the blacksmith shop does not take in such a wide range as that used in a large machine shop, as steel of 75 points carbon will answer all purposes in the blacksmith shop, but in the machine shop steel is used of a much higher carbon, ranging up to 100 points or 1 per cent and even higher. High carbon steel is used chiefly for making lathe and planer tools, which has been found out by practical experience to be preferable owing greatly to the reason, that these tools do their work by steady pressure. Should a cold chisel be made from high carbon steel, say 1 per cent, the head of the chisel would be continually breaking and splitting off. High carbon steel is more difficult to weld and will harden at a less heat than low carbon steel. Tools that are to do their work by striking with a hammer, as a cold chisel, should always be made from a medium low carbon steel. But in these days, steel can be had in any shape or temper to suit any tool, so when ordering steel from the manufacturer always state what kind of tools the steel is to be used for. The percentage of carbon which the steel should contain for different tools will be fully explained throughout the book, as each tool is described.

The Cracking of Tools When Hardening and the Cause.

The cracking of tools during the hardening process, is one of the great obstacles to be overcome by the steelworker, and which is the cause of the loss of a great amount of expensive tools and labor.

The primary and main cause for tools cracking when hardening, is overheating of the steel, another cause is by uneven heating, still another cause will result from forging and leaving strains in the steel by irregular heating and hammering, and also by improper annealing. Steel that is heated in the blacksmith's fire is very liable to crack in hardening, unless great care be exercised, and the tendency for the steel to crack will be increased, if the mechanic has only a limited amount of knowledge as regards the nature and virtue of steel. For illustration I will give the way that a great many who call themselves steelworkers, harden a piece of steel. They will take the piece of steel, place it in the fire, then turn on or blow a very strong blast so as to heat it quickly, getting one part at a white heat while another part is barely red, then plunge it into the water any way to cool off, consequently it cracks, and the operator blames the steel saying it was no good, while he himself was to blame. Laying the blame on the steel is the theory of a great many blacksmiths and steelworkers, especially when the tool does not give good satisfaction. But on the other hand, should the tool do good work they are ready to take all the praise to themselves by telling others about it, and I state this from experience as I have been in the same position before finding out my mistake.

Now let us harden a piece of steel properly as it should be done, and for example, supposing we have a flat piece to be hardened, 2 inches square by ½ inch thick. Place the steel flatways on top of the fire, heat slowly and very evenly, turning the steel over occasionally so as not to heat in streaks, until the whole piece becomes heated to a cherry red or just enough to harden, then cool edgeways from an upright position, in clean cold water allowing it to remain there until it is perfectly cold, and it will be perfectly hardened and free from all cracks.

And to more fully illustrate, I will relate a little incident in my own experience. I took a position as toolsmith in one of the large shops of the Chicago, Rock Island and Pacific Railroad. The first job I undertook to do, was to harden and temper a great number of flat thread cutting dies, as I started to work the machinist foreman came along and said to me, "I want you to harden these dies without leaving cracks in them." A few days after the dies had been in use, I asked him if he found any cracks in the dies, and he replied "no, not one." Then he went on and explained to me, that the toolsmith who was there before me, was continually leaving cracks in the dies and laying the blame on the steel saying it was no good, while the dies were not giving good satisfaction and at the same time keeping a machinist busy making new ones and keeping others in repair. A few days later the machinist (who was keeping the tools in repair) came along to me and said, "I am not working much more than half my time since you started, as I have not near so many dies to keep in repair." Reader, I have not related this affair to give myself praise, but instead, to point out to you the difference between two mechanics and both calling themselves toolsmiths, one having very little knowledge concerning the nature and working of steel, was giving poor satisfaction, spoiling a great many tools and resulting in the loss of his position.

The other had a thorough knowledge of steel in every way and did his work in a highly satisfactory manner.

The first man did his work by heating his tools too fast, having one part of a tool at a white heat and another part scarcely red, and when being cooled to harden they cracked. The second man did his work by watching carefully so as to heat the tool very evenly, no part of the tool being any hotter than just enough to harden, the results being every tool came through the hardening process safe and sound, without a flaw of any kind. Reader, which of these mechanics are you going to be, first or second? Consider the difference between the two, then take your choice.

How to Judge Hard from Soft Steel.

There are numerous ways of telling the difference between hard and soft steel, as in the following. First way is by the fracture of a fresh break, as hard steel when broken cold from the bar, will show a very fine and smooth fracture, while soft steel will show a coarse and rough fracture. Second way, take two bars of 3/4 inch octagon steel nick the bars all around when cold, 6 inches apart for cold chisels, place the part of the bar at the nick directly over the square hole of the anvil. Then strike with a sledge. The hard bar will break at the nicks with one or perhaps two blows, but the soft bar will require five or six blows in order to break it.

Third way, supposing a number of cold chisels are to be dressed that have been in use for some time, by close observation it will be seen that the heads are of a different shape and appearance. For instance. The head of one chisel will have the steel widened out and curled down over the body of the chisel. This illustrates soft steel of about 60 points carbon. Another head will crumble off as it widens out instead of curling up. This indicates steel of medium hardness of 75 or 80 points carbon, which is the best for cold chisels and all similar tools. Still another head will show the steel split and broken off half an inch down the sides. This represents a high carbon steel of 1 per cent, which is too hard for chisel use, but would be good for lathe and planer tools. This way is perfectly reliable when telling the difference between hard and soft steel. When forging, soft steel will give much more readily under the hammer and will hold the heat much longer, than hard steel.

If two pieces of steel, one hard the other soft, are hardened at the same heat, the hard piece will be scaled off white, while the soft piece will be only partly scaled off, showing a black and white appearance.

How to Tell Good from Poor Steel.

The fracture of good steel when first broken, will show a silvery white appearance clear through the bar, while the fracture of poor steel will show a dull grey. When judging or testing steel by the fracture do not allow the steel to get wet or rusty.

Testing Steel After Hardening With a File.

When hardening tools of an expensive nature, it is always best to make sure the steel is properly hardened before undertaking to draw the temper, by testing with a good sharp file in some part that will not interfere with the cutting qualities of the tool. Should the file run over the steel without taking a hold, the steel is all right, but on the other hand, should the file take a hold of it, the tool will have to be hardened again, having a little higher heat than the first time.

Instructions on Toolmaking That Have to Be Given Many Times.

In giving instructions on toolmaking in the following chapters of this book it will be necessary to give to a certain extent the same advice as different tools are described. And so I will ask the reader to bear this in mind, as what is told many times will be that which is most beneficial towards making steel work a success, and also which I wish to impress most deeply on the reader's mind in order that it may be well remembered.

CHAPTER II.

The blacksmith's fire—Anvil—Tongs—Making a hand hammer.

The Blacksmith's Fire.

The fire is one of the most important things connected with the blacksmith's or tooldresser's trade, and is the first thing I will describe toward toolmaking. The main points of a fire to be considered is, the fire should be on a forge large enough to enable the fire to be easily regulated to any size, according to the work that has to be heated in it, and have plenty of blast which can be well regulated. The fire should always consist of well charred coal, being perfectly free from all sulphur, gas, ashes, clinkers and thick smoke before undertaking to heat steel in it.

In reference to the size of the fire I will illustrate, supposing we have a large piece of steel to heat (say a stone hammer), we want to heat it evenly and clear through, the fire must be large enough to accommodate the hammer so that it will not come in contact with the blast from the blower or bellows, and still have a certain amount of fire over the hammer, which will require a fire of about 6 inches deep and 8 inches across the surface, but a smaller fire will do in case of small tools.

The author has used a fire that was barely $1\frac{1}{2}$ inches across the whole heated surface, but this was made especially for hardening and tempering certain parts of tools. But be on the safe side by having the fire large enough, as coal is cheaper than steel and saves time.

It will be money saved by securing as good coal as is possible to get, there being a great difference in coal, as some kinds are more free from smoke and sulphur and will not cake or get hard as other kinds, and tools of irregular shape can be placed more easily without disturbing the build of the fire. Keep the coal under cover and clear from all rubbish, coal loses a great amount of its heating qualities when the sun shines on it continually.

Bellows and Blowers.

In the majority of ordinary shops the bellows are still used, some being much better than others, both as to the power of the blast and the construction of the bellows, and to anyone who is following the trade of blacksmithing I would advise having as good a one as is possible to get.

I have seen bellows that have been in use for fifteen years and are almost as good as new, while others will wear out in one year. The hardness or easiness as the case may be of blowing the bellows is chiefly due to the way they are set up (as I have seen bellows that would tire a blacksmith to blow them, which should be in the nature of a rest instead of hard work), and if the uprights or posts on which the bellows hang are in a very upright position, the chances are the bellows will be hard to blow, so set them at an angle of about 45 degrees. If the uprights are given too much angle there will be too much leverage, and the bellows will lack motion.

Blowers are used a great deal in these days (they take the place of bellows), which are run by hand in small shops by the use of a crank or lever, although in large shops blowers are run by steam power, and the smith simply regulates his fire by moving a lever to different positions. It makes no difference what produces the blast so long as there is plenty of it, and at the same time it can be well regulated.

The Anvil.

The anvil is a tool used in the blacksmith trade or shop which is practically the foundation of all tools, for the forging and shaping of all classes of work and more particular in regards to toolmaking. There are a great many different sizes of anvils, as well as a great many different makes. In reference to the size of the anvil, some smiths want one size, some another, but for general tool work an anvil weighing 225 pounds is about right.

There is a great difference in the make and also the quality of anvils. The author has forged tools on almost every make of anvil manufactured in Great Britain or the United States, but the kind that has given the most satisfactory results, and which can be relied upon, is known as the Hay Budden. Manufactured by the Hay Budden Manufacturing Company, Brooklyn, New York, U. S. A.

This make of anvils is fully guaranteed, they are made from the very best material and by expert workmen, and the face is perfectly hardened. There are no soft spots, neither do they sag down or get hollowing in the face, as in a great many other makes, including the Peter Wright. This is no hearsay neither is the author favoring any particular manufacturer, but this advice is founded on experience which is for the reader's benefit, as in order to do good work a good anvil is necessary, which is perfectly free from all hollows and soft spots. Tools of a flat surface, such as cold chisels, mill picks, axes, etc., must be dressed on a hard and smooth anvil face to obtain the best results. When dressing wide tools such as an axe, a rough faced anvil will produce strains in the steel, which will increase the tendency to crack when hardening.

The height of the anvil when on the block depends upon the tallness of the blacksmith who is to forge on it. I have noticed cases where a blacksmith was working over an anvil so high as to be unable to strike a good hard blow, while on the other hand a tall blacksmith was working over a low anvil that was making him humpbacked and round shouldered. But a good rule to go by, which will be found about right for all, is to have the anvil just high enough so that the blacksmith may readily touch the anvil face with his knuckles when clasping the hammer handle and standing in an upright position.

Don't have the anvil merely sitting on a block that is continually jumping up and down with every blow from the hammer, but have it well bound to the block. But some will say, "That stops it from ringing," or "I can't work on an anvil that does not ring." Reader, this is all nonsense, what has the ringing of the anvil to do with the work, it may be all right for the class that wants to make a lot of noise to let others know they are working, but it is of no use when it comes to doing the work with ease, both for the blacksmith and helper. So have a block a little larger than the base of the anvil, a good depth in the ground, say 3½ or 4 feet (if it can't be made solid any other way), place the anvil on it and bore a hole through the block 6 inches below the anvil.

Now make a bolt from ⅝ round iron 2½ inches longer than the block, as shown in Figure 2 at **a**. Make

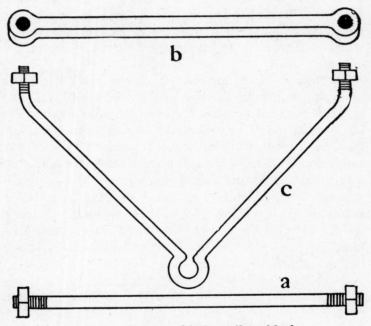

Fig. 2. Irons to bind anvil to block.

two irons from ⅝ square as **b**, leaving them long enough so that the holes in the ends will extend over the base of the anvil. After taking the measure make two irons, as **c**, which are to extend from below the bolt in the post up through the irons that rest on the **base** of the anvil **b**. Place all together and tighten up the nuts firmly. Figure 3 illustrates the **anvil fast**ened to the block. In large up to date shops cast iron blocks are used, which are made especially for the anvil to fit into, thereby holding the anvil perfectly fast.

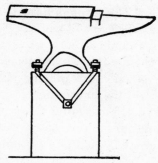

Fig. 3. Showing anvil bound to block.

Tongs.

Tongs are among the most necessary tools needed by the blacksmith, and without them he would be at a standstill. There are a great many different shapes and forms of tongs (with the exception of a few ordinary styles) which are made according to the work they are to hold, and to be a good tong maker is an art to be proud of, as it requires skill to make them light and strong, and have them hold perfectly fast to the work without hurting or cramping the hand.

Success often depends on good tongs, as I have known blacksmiths to fail at their work simply for the want of them. Again, a blacksmith who uses all his strength to hold clumsy or poor tongs on the work cannot use his hammer to advantage. The author has tongs that are made entirely from steel and are very strong, and also are only half the weight of the ordinary iron tongs. They are a pleasure to use, as they hold perfectly firm with very little hard pressure, because the handles just come close enough together so as to keep the hand full, in somewhat the same way as clasping a hammer handle. These tongs have been in use a number of years and are as good as new.

In using tongs made from steel care should be exercised to keep them cool (and especially if made from cast steel) by occasionally placing them in water. If brine is used to do the hardening keep the tongs out of it or they will rust, but should there be occasion to put the tongs in it be sure and rinse off after in clear water.

One of the greatest features in making tongs at the present time is to have them adjustable, so that one pair will do the work of seven or eight ordinary pairs of tongs. I have tongs in my possession that will hold from ½ inch square up to 1½ inches square, or octagon, and will also hold flat sizes 2 inches by ⅝ down to ½ square, and hold each size very firm, while the size is regulated by an adjustable jaw by moving a small bolt into different holes in the jaw, which requires but a few seconds to change, as shown in Figure 4. Still another point worthy of mention is to make the handles half round. This will form a spring and will be very easy to hold, but have the half round as wide as is reasonable; for example, ⅝ half round will be right and fill the hand better than 7-16 round. Figure 5 illustrates tongs specially adapted for dressing cold chisels, made with V-shaped jaws, which will readily hold round, square or octagon. When making tongs, as Figures 4 and 5, from cast steel use a very soft steel of about 60 points carbon, and ⅞ square in size. After forging the jaws, as shown in Figure 5, the handles may be

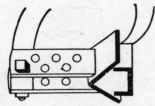

End view of adjustable tongs when holding smallest size.

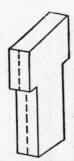

Showing how adjustable jaw is forged—dotted lines indicate where to split.

Showing piece split and forged to shape of jaw.

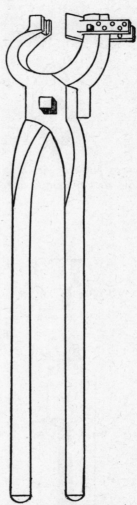

Fig. 4. Adjustable tongs regulated for holding largest size.

Figure showing jaws made to hold round octagon or square.

Illustrating how jaws are forged and bent to shape.

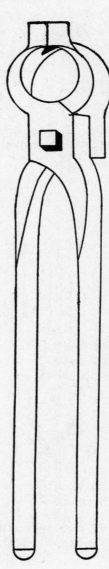

Fig. 5. Tongs specially made for dressing old chisels and all similar tools.

forged or welded on, but will prove most satisfactory if forged in one piece.

Figure 6 illustrates double levered and adjustable tongs invented by the celebrated steel worker, Prof. W. S. Casterlin, Pittston, Pennsylvania, U. S. A. These tongs have been improved by the author and are very powerful and light. They are especially designed for grasping tools of a beveled nature, such as mill picks, axes, etc., and will hold flat material of any width, ranging in thickness from 1-16 to 1¾ inches. The jaws of these tongs are copper lined, which prevents slipping.

Fullers and Swages.

These tools are invaluable to the blacksmith in shaping and forging many classes of work, especially in toolmaking. Fullers and swages take in a wide range of different sizes, from ¼ to 2 inches for ordinary use,

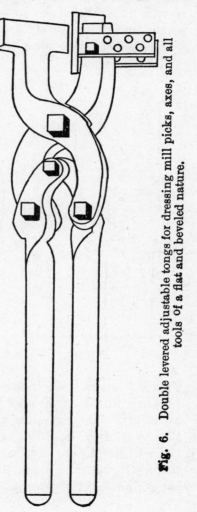

Fig. 6. Double levered adjustable tongs for dressing mill picks, axes, and all tools of a flat and beveled nature.

and larger sizes are used according to the class of work to be done. When making fullers or swages use a soft carbon steel of about 65 points. They will not require to be hardened or tempered as their work is chiefly on hot iron or steel.

Figure 6½ represents a top and bottom swage, while

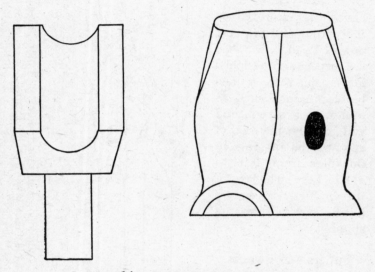

Fig. 6½. Top and bottom swages.

Figure 7 shows a small size of top and bottom fullers. Figure 8 illustrates a large top fuller. Figure 9 indicates how large swages, and also large fullers, as Figure 8, may be forged before bending to shape. Dotted lines as **aa,** illustrate the projections, **cc,** bent to form a fuller. Dotted lines, as **bb,** illustrates the projections bent to form a swage.

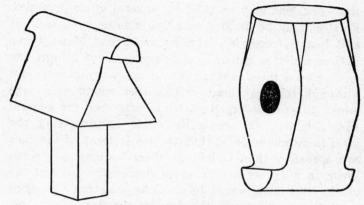

Fig. 7. Top and bottom fullers.

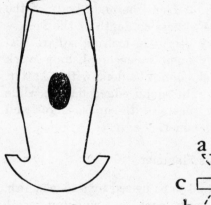

Fig. 8. Front view of large top fuller.

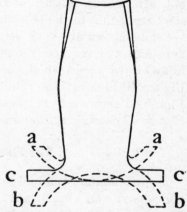

Fig. 9. How large fullers and swages may be forged.

Flatters and Set Hammer.

The flatter is to the blacksmith what the plane is to the carpenter, being principally used for taking out all hammer marks and so leave a finished appearance on the work, but may be used for several other purposes. Flatters may be divided into two classes, such as light and heavy (most blacksmiths make use of only one flatter, which is generally a heavy one, but a light one can be used to advantage in a great many cases), and although they are made with either round or square edges, the round edged flatter is preferred for general use. Flatters are generally made by upsetting the steel to form the face, then the projections of the face are spread with a fuller or otherwise may be driven down in a square socket, same size as the body of the steel. But flatters may be made by selecting a piece of steel the same size as the face of the flatter is to be, then fuller in from all four sides of the steel and draw out, afterwards cutting off from the main body of the steel according to the thickness or depth of the face.

Set hammers are very useful in making square corners and are very convenient in accomplishing work in awkward places which cannot be done with a flatter. Figure 10 illustrates a light round edged flatter, while Figure 11 shows a heavy flatter with square edges, and Figure 12 shows a set hammer.

The Hammer.

A good hammer is a tool to be prized by a blacksmith, and plays an important part in working steel. The face of the hammer must be properly hardened and tempered, in order to prevent it from getting hollow in the center or being too hard on the corners, thus causing it to break, and the face must be perfectly smooth so as not to leave any nicks or dents in the steel.

A great many blacksmiths think all they have to do

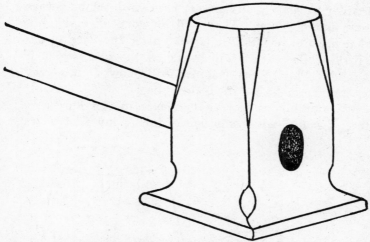

Fig. 10. Light round edge flatter, made from 1¼ inch square steel, face 2 inches square.

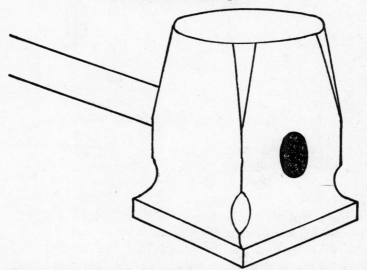

Fig. 11. Heavy square edge flatter made from 1½ inch square steel, face 2¾ inch square.

in order to get a good hammer is to go to the hardware store and buy one. They will buy one all right, but what are the results. It is this, after they are in use a little while a piece will break off the face, which if examined closely the fracture will show a dull gray appearance, a sure sign of poor steel.

By referring to buying hammers in the hardware store, I had a wide experience during my first years

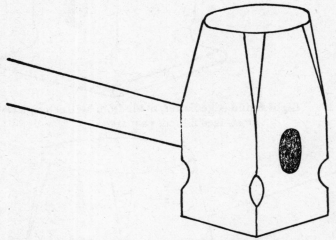

Fig. 12. Set hammer made from 1¼ inch square steel.

at the trade, and through ignorance, after purchasing a hammer and it did not give good satisfaction, I would generally give the hardware man a calling down for selling it to me. Although he was not to blame as his business was to sell and not to manufacture, and as regards his knowledge concerning good or bad steel, it was very limited. I am not stating this, reader, to criticise the hardware merchant, or the tool manufacturer, or machine made tools of any kind, as in a great many cases the machine tools are equal to those that are hand made. But not so in the case of a good hand forged, hardened and tempered hammer.

Making and Dressing a Hand Hammer.

In making a hammer, as to weight and shape, it is hard to say what would suit all blacksmiths, but for a forging hammer for making and dressing tools, two pounds in weight will be about right. But in cases such as dressing mill picks, granite tool sharpening, and on jobs of tool dressing where a helper is not necessary, a hammer of three and one-quarter pounds in weight will be best.

But let us take a piece of good steel of about 75 point carbon, 1½ inches square and 4 inches long, which will make a forging hammer two pounds in weight. Have a good clean fire and plenty large enough; place the steel in the fire and heat slowly, turning it around occasionally so as not to overheat the corners while the center is yet black, but heat to a good even yellow heat clear through. The hammer may now be forged as shown and illustrated in **a b c**, Figure 13, or any shape the blacksmith may choose. I have known blacksmiths who could forge a well shaped hammer equal to the expert toolsmith, but when they came to the hardening and tempering of it would fail entirely.

I will state some of the obstacles that blacksmiths have to contend with which reduce the chances of making a good hammer. They are afraid to heat the steel to a high forging heat; and to explain, I remember when in my apprenticeship I was helping a blacksmith to make a hammer, and as we were proceeding he was giving me instructions while I was blowing the

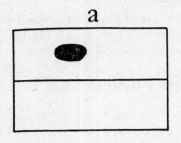

Piece of 1½ inch square steel 4 inches long, with eye punched to make a 2 lb. forging hammer.

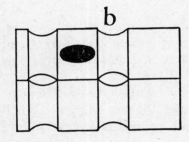

Illustrating how fullering is done after eye is punched.

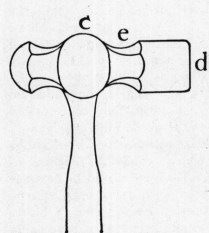

Showing side view of hammer completed with eye spread at c, (which is done with a fuller while the eye pin is in the eye of the hammer) and fullered in at e, with a larger fuller at the finish.

Fig. 13. Indicating how hammer is forged.

bellows, he said: "Never heat a piece of steel hot enough to raise a scale," (and of course at that time I thought the advice was good), however, we worked away, getting the eye punched, which was quite a hard job at that low heat, considering we had a 2 inch square piece of steel, but while we were making the hammer we had other jobs to do as is generally the case in a general blacksmith shop, and so occasionally the steel was left soaking in the fire for half an hour at a time (letting the steel lay in the fire is a bad practice among blacksmiths and is very injurious to the steel after being hot enough to work). Well, we got it forged to a very good shape, but I was not there when he hardened and tempered it (as I had given up my job). However, I called in some time after and noticed that the hammer was broken off at the corners, and also being hollow in the face.

Another blacksmith who had forged a well shaped hammer and after hardening and tempering it in his own way, found it to be as soft as lead. This hammer, I think, was hardened and tempered three or four times without success, consequently he threw it in the scrap pile, saying the steel was no good, while he "himself" was to blame as he had a very poor knowledge pertaining to the working of steel.

Now I would like to impress deeply on the reader's mind, if every mechanic could be an expert by doing his work in a rough and tumble way, the world would be full of expert mechanics, and if every blacksmith or toolsmith could make tools to give unlimited satisfaction, there would be no use in writing this book. But as this is impossible, it is necessary to have a thorough knowledge of working steel when forging hammers, and moderation in heating is the stepping stone to success.

For instance, if a square piece of steel is heated very hot and fast the corners will become overheated and if struck a blow with the hammer on the corners the steel would fly to pieces, while the steel has lost its good qualities and is spoiled in the beginning. The other extreme is trying to forge steel at a low heat, sometimes not above a blood red, while every blow that is struck on the steel is putting strains in it. What I mean by "strains" is the steel must be hot enough so that it will be worked clear through, and if this is not done the steel is liable to crack in the hardening. I have known hammers to crack clear through the center of the face, then all the labor is lost.

But to get the best results, forge the hammer in the beginning by having the steel at an even yellow heat, but lower the heat as the hammer is finished (all tools should be finished at a low heat, for example, a dark dull red, in order to get the best results). If much filing is to be done after the hammer is forged it should be annealed. When dressing the face end of the hammer make the corners a little rounding, but otherwise have the face perfectly flat and level, and the hammer is ready for the final blow, "the hardening."

Hardening and Tempering a Hammer.

After hardening there are different ways of tempering a hammer according to the shape. But for illustration, I will harden and temper a hammer as in Figure 13, but before commencing it must be understood that the face end of the hammer, at **b**, is the principal end. After getting the fire in good shape, the coal well charred and free from all sulphur and thick smoke, place the face end of the hammer in the fire from an upright position; now heat slowly and evenly, making sure that the corners do not get overheated; should the corners get hot enough to harden before the center, stop blowing the fire until the center has come up to a cherry red, or hot enough to harden even with the corners; then plunge the whole hammer in the hardening bath and hold there until stone cold. Now polish the face end (that was hardened) bright, then place the round or pene end of the hammer in the fire in an upright position and heat very slowly, as while heating to harden the round end the temper will draw to a blue in the face end if properly timed or regulated, so that one end may be hardened and the opposite end tempered in one operation, but be careful to watch both ends of the hammer at the same time. If the round end gets hot enough to harden before the temper appears on the other end stop blowing the fire until the temper begins to show up. It will not matter about the round end, after hardening, whether any temper be drawn or not, as there are no sharp corners to break off, but the temper may be drawn by holding over the fire and keep turning the hammer around, or may be done by placing a heated heavy iron band over the end.

Bear in mind that when heating the ends of the hammer to harden do not heat to a hardening heat more than ¾ of an inch back from the end, and at all times never harden the eye. A hammer that is made, hardened and tempered after these instructions will not get hollow in the center or break off at the corners or the eye. This is the author's favorite way of hardening and tempering a forging hammer and will be found very simple by the average blacksmith when once tried. Although there is another good way which is very convenient for hammers that have two face ends, such as a horseshoer's turning sledge or a boilermaker's hammer. For example, after the ends are hardened the temper can be drawn by placing a heated heavy iron band over the end of the hammer (as already mentioned). This method will draw the corners to any temper (say a pale blue), while the center will remain hard. This way will give good satisfaction as the center of a hammer cannot be too hard unless overheated, but the corners must be drawn to a very low temper. These instructions will apply to all ordinary hammers and sledges.

The old-fashioned way of tempering a hammer is by heating one end with the corners hot enough to harden, while the center is barely red, it is then cooled off by dipping an inch into the hardening bath, then allowing the temper to run down to the desired color. Consequently the hammer becomes hollow faced, could not be otherwise, as it did not harden in the center of the face at all, because it was not hot enough, and when the temper came down it made it still softer. For example, supposing a hammer face (unless it be a very small hammer) is heated very evenly, then hardened by being dipped an inch into the water, it is still bound to be soft as the temper is sure to run out at the center first.

Successful Points to Be Remembered in Making and Tempering a Hammer.

Have the eye straight through the hammer and a little smaller in the center, which will keep the handle in the hammer much better after being well fitted and then wedged. Always have the center of the hammer face as hard as the corners, but the corners must not exceed a light blue temper, unless the hammer is forged from a very soft cast steel, harden the hammer at as low heat as it will be sure to harden at. Have a good straight handle, a little spring to it is a good fault. Another good point to remember when making hammers is, do not punch the eye the full size at the beginning, but have it a little smaller, as the eye will get larger as the hammer is forged.

Punching Holes in Steel.

Punching holes in steel is considered by the average blacksmiths to be a difficult job, the trouble being they try to punch the steel at too low a heat or else they have not proper tools to do it with. Some blacksmiths use too long and straight an eye punch, consequently when the punch enters the steel a short distance the end becomes hot and upsets, causing it to stick in the steel, while the blacksmith experiences a difficulty to get the punch out of the steel. Then the punch is straightened again and the blacksmith works away until he gets the hole through. I have known blacksmiths to have a punch stick in the steel three or four times while punching one hole through a piece of 1½ inch square steel.

Coal dust is very good when put in the hole to keep the punch from sticking, but the main points to be considered is the heat in the steel when punching and proper shaped tools, especially when making hammers. And so make an eye punch and eye pin, as Figures 14 and 15. The eye punch is made from 1½ inch square steel. After the eye is punched the punch part of the tool should be forged down to an oval shape diagonally across the steel (which will bring the handle at a right angle when the punch is in use and also keep the hand away from the hot steel), and be very short and have plenty of taper, while the corners of the extreme point

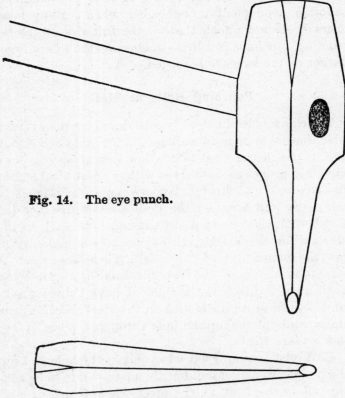

Fig. 14. The eye punch.

Fig. 15. The eye pin.

should be perfectly square, an eye punch of this description will not stick in the hole and will not bend.

When punching have the steel a high yellow heat, then the punch will penetrate it with more ease than if heated to barely a cherry red, and cool off the punch occasionally. An eye pin should be made from $7/8$ square steel, by forging it tapering to a small square point, then to make oval, hammer down two opposite corners of the square a little rounding, which will give the shape required. The eye pin should also be short and have plenty of taper, this will make the eye a little smaller in the center when driven in from opposite sides, which will keep a handle in much better if well fitted in then wedged.

CHAPTER III.

The cold chisel—The hardy—Drills and drilling.

The Cold Chisel.

A cold chisel is a tool used by every mechanical trade and business where iron or steel is used, or wherever machinery is repaired or manufactured, and in the proper forging, hardening and tempering of a cold chisel lies the foundation and successful stepping stone in making all edged tools with a flat surface, as a cold chisel hold a good cutting edge and neither bend or break.

But the shape of the chisel is another point that must be well understood, as a fine chipping chisel which is made very thin for use on solid and soft material would not do for a boilermaker who is working on sheet steel which vibrates at every blow from the hammer. Thus, the vibration of the steel would be very trying on a thin chisel, and would consequently cause it to break, and so a heavier and thicker chisel must be made and put into use, as shown in 1, 2, 3, 4, Figure 16.

A cold chisel can be made to chip almost any kind of material, as the author has made chisels to chip from the softest known material up to chilled metal, which will seem like a fable to a great many simply because they never saw it accomplished. I have known machinists who wore out a number of new files performing some work on chilled metal, because they could not get a cold chisel properly made to chip it.

Now I would like to impress on the reader's mind that I have made chisels to chip metal that a file would not bite, but these chisels were not tempered to a blue (as a great many mechanics think a cold chisel must always be tempered to a blue), neither were they made from any old scrap piece of steel that might be handy. As I have known blacksmiths to take an old file or rasp, forge it to a round or square, and then attempt to make

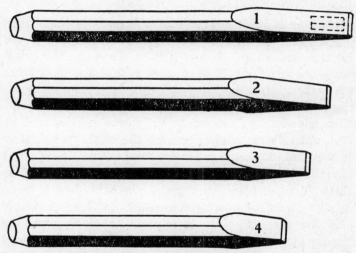

Fig. 16. Illustrating shapes of cold chisels according to use. 1. Machinist's chipping chisel. 2. Ordinary or farmer's chisel. 3. Boilermaker's chisel. 4. Chisel for chipping hard metal.

a cold chisel out of it, what nonsense. File steel as a rule is too high in carbon, being 1 per cent and over, while 75 point carbon is plenty high enough for cold chisels. Again, files in a great many cases are manufactured from a poor grade of steel; not only that, but the cuts or teeth of the file will be put deeper in the steel the more it is forged, consequently as the tool is finished the teeth of the file will be in the cutting edge of the tool (which is to be used as a cold chisel), which will break out or upset when put in use.

I will explain how the majority of blacksmith's harden and temper a cold chisel. After they have it forged to the shape, finishing sometimes with a low black heat, other times at a high yellow heat, no matter whether the last blows fall on the edge or the flat side, whichever is most convenient to bring the chisel into the desired shape, then file on the cutting edge, now for the hardening and tempering. They get the chisel hot anywhere from ¼ of an inch to 2 inches back from the cutting edge, dip into the water ¾ of an inch, holding it there for a minute or so, brighten it up a little, then allowing the temper to run down, no matter how fast, until a blue temper reaches the cutting edge, then it is cooled off and is ready for trial. Now I will point out the dangerous practices that the chisel has come through, when being hardened. The chisel was lowered into the water about ¾ of an inch and held perfectly still for a short while, now right there between the hardened and the unhardened (if the chisel was heated enough to harden far enough back) is a dividing line and a strain, and a great many chisels are broken off there from the cause of this strain. Again, if there was plenty of heat left in the steel above the hardened part, especially if hardened barely 1 inch back from the edge, the temper will run down very quickly, so that when it reaches the cutting edge, there is only ¼ of an inch that shows any temper which will be a blue, the other colors have hardly been noticed on account of the temper running down so rapidly. Thus there is only ¼ of an inch back from the cutting edge of the chisel that is tempered, while back of the tempered part the steel will be extremely soft, which is apt to cause the cutting edge to bend or break off and should the chisel be thin the tendency to break or bend will be increased.

But let us make a chisel, as No. 1, Figure 16, that will when finished cut the bar it is made from and not bend or break if used any way like a chisel should be used. Take a bar of $7/8$ inch octagon good steel about 75 point carbon, cut off 6 inches, which will make a chisel about the right length. The end that the hammer is to strike upon should be drawn down a little and left square or flat on the end, then the blow will fall directly on the center of the body of the chisel and be less liable to break it. Heat the steel for the chisel end to a good yellow forging heat, draw it down near to the required shape, making it a point to have it a little narrower than just what is wanted when finished, and we will finish the chisel in a way not known to the ordinary blacksmith. Now have the chisel a low red heat so that it may be noticed to be red when placed in a dark place. Then strike the chisel five or six good hard blows on the flat side (holding the chisel very firm and level on the anvil), commencing about 2 inches back from the cutting edge, then coming gradually towards the cutting edge with every blow, when the last blow has fallen directly on the cutting edge turn the chisel over and hammer this side the same way, but be careful and do not hammer the steel too cold, but instead heat again and hammer both sides evenly as before. Then it is finished and ready for the hardening.

The hammering of the chisel on the flat sides when at a low heat refines and packs the steel, leaving it dense and much stronger than steel in its natural state, but remember this, not one blow is to be struck on the edge of the chisel, as it would knock out all the tenacity and toughness put in by the blows on the flat side, if the edges spread out a little uneven during the hammering grind or file to the right shape. Hammering steel after it gets cold or below a certain heat is injurious and makes the steel brittle and flaky as pie crust, and will never hold a good cutting edge until cut off. When hammering wide chisels, hammer in the form of dotted lines in chisel No. 1, commencing as indicated, and on both sides equally. Now we will harden and temper the chisel, after being forged as already explained, heat the chisel to an even cherry red heat 1¼ inches back from the cutting edge, then dip deep in the hardening bath at least 1½ inches, at the same time raising and lowering so as to form no dividing line of strain (and thus the hardened part will soften away gradually into the unhardened part). Polish up bright with some sand paper, emery cloth or polishing stick, as mentioned in the following pages, but after polishing there may not be enough heat left in the chisel to drive the temper down to the cutting edge, and so draw the temper by holding the chisel over the fire, heating very slowly and moving back and forward so as to get an even light blue temper all over the hardened part, then cool off. A chisel that is forged, hammered, hardened and tempered after these instructions will give the best of satisfaction, and can be placed over the horn of the anvil and struck quite a hard blow with a hammer, flatways on the chisel, without danger of it breaking or bending. The reason the chisel stands this abuse is due to the heavy hammering that was done on the flat side at a low heat, and also by being properly hardened and tempered so far back acts as a foundation back from the cutting edge.

The chisel, No. 1, Figure 16, is very thin being 1-16 thick at the cutting edge and tapering back 2½ inches to ½ inch in thickness (when made from ⅞ steel), and is classed as a machinist's chipping chisel. It will do for almost any machinist's ordinary work, but is too soft to cut anything harder than cast steel. A chisel to chip hard chilled metal, especially if struck with a heavy hammer, must be made with a short taper to form the cutting edge, as No. 4, Figure 16, and is tempered to a purple. If this is found to be too soft use a harder temper, say, a copper or dark straw. But if the metal is excessively hard, harden and draw no temper. Chisels that are made thick, are not required to be hardened and tempered so far back as thin chisels. This information applies to all kinds of chisels and similar tools. Nos. 1 and 2, Figure 17, illustrates the cape and round nose chisels. When making these chisels they should have a clearance by making the cutting edge, as A, No. 1, wider than at b, the round nose is forged round on the under side as c, by being placed in a bottom swage.

The Hardy.

The hardy is a tool used by almost every blacksmith, but more especially by general blacksmiths. It is used for numerous purposes, but its main use is, as a chisel. The cutting edge should be made thin, as the hardy is chiefly used for cutting hot iron or steel, and again if made thin it will not require so many blows from the hammer to do the work. The forging, hardening and

62   THE TWENTIETH CENTURY

tempering are the same as mentioned in making a cold chisel. But the points to be considered when making one is, have the shank that fits into the square hole of the anvil, fit snug so that it will not twist round, while the upper part should extend well over the sides and should be level, so as to sit on the anvil solid.

Sometimes it is necessary to make a hardy to go in the round hole of the anvil. A hardy of this kind, is made by splitting the steel, so that one part will extend over the side of the anvil which will keep it in place, as shown in Figure 19 at a. Hardies of this

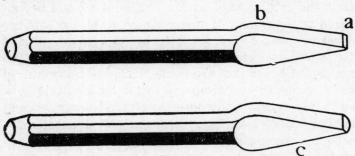

Fig. 17.   Ordinary shapes of cape and round nose chisels.
1. The cape chisel.   2. The round nose chisel.

shape are used chiefly by toolsmiths when sharpening stone cutters' tools, as the square hole is occupied by holding other tools such as a stake which remains stationary. And as a hardy is also very needful, it must be made to sit in the round hole, which is the only remedy.

Heavy, Hot, Cold, and Railroad Chisels.

I have known a great many blacksmiths to try and do all their work with one chisel. Now any reasonable thinking mechanic will know that a cold chisel will

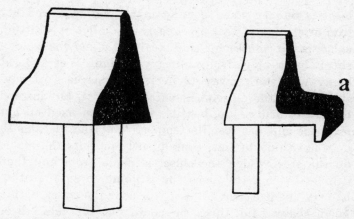

Fig. 18. Correct shape of hardy.

Fig. 19. Hardy made to sit in round hole of anvil.

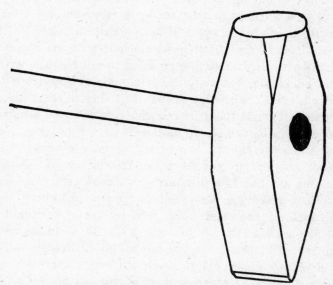

Fig. 20. The blacksmith's cold chisel.

lose its temper when coming in contact with red hot iron or steel, and that a properly made hot or splitting chisel is too thin to cut cold steel and stand the sudden shock by a blow from a heavy sledge. Yet a great many will work away trying to cut a piece of cold steel, while the chisel will barely mark it, but instead the chisel bruises up, because it has been used on hot material and has lost its temper. But the blacksmith does not think of this, consequently he puts the blame on the steel which the chisel is made from and then decides to heat the piece (he is going to cut), so that he may cut it easier. But it is a poor theory as with a good heavy cold chisel he could cut off cold, three pieces (up to a certain size), while he is heating one piece. But a great many blacksmiths think it is not possible to cut off tool steel without heating it.

But to impress on the reader's mind the necessary use of good chisels, we will take a railroad chisel, if it is well made and from good steel, it will cut at least 15 rails, while I have known some to cut 60 rails without redressing, but they did not come in contact with hot material. Although there are today a great many blacksmiths and tooldressers who cannot make a chisel cut two rails, which applies also to the blacksmiths' hot and cold chisels, a good blacksmith's heavy cold chisel should cut off at least 120 pieces of $7/8$-inch octagon steel of 75 point carbon without grinding.

When making a blacksmith's hot or cold chisel, use $1\frac{1}{2}$ inch square steel which will be heavy enough for all ordinary work. A hot chisel should be made very thin so as to penetrate the hot metal with ease, but a cold chisel must be left much thicker. A hot chisel, when doing a great amount of cutting at once, should be cooled off occasionally to avoid drawing the temper as little as possible. These chisels are forged, hardened and tempered as explained in making an ordinary small cold chisel, but heavy chisels, such as railroad chisels, instead of hammering with a hand hammer at the finish when forging, place a heavy flatter on the flat side of the chisel and let the helper strike it a few good heavy

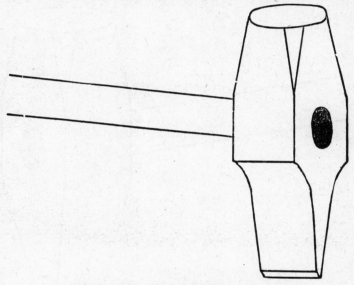

Fig. 21. The hot or splitting chisel.

blows with a sledge, then turn over the chisel and go over this side the same way, which will have the same effect on the steel as hammering with a hand hammer, and be better, as the blows will be much heavier, but be sure that the flatter comes in contact with the extreme cutting edge.

A railroad chisel is made from 1¾-inch square steel. It should not be made entirely square across the cutting edge as an ordinary chisel, but should be ground or filed rounding at the corners, as in Figure 22, which will make them more difficult to break and also give better satisfaction. If the cutting edge is tapered in from the sides to 1⅛ inches across the cutting edge, it will cut deeper when struck with a sledge or hammer and will be more convenient to get in the corners of

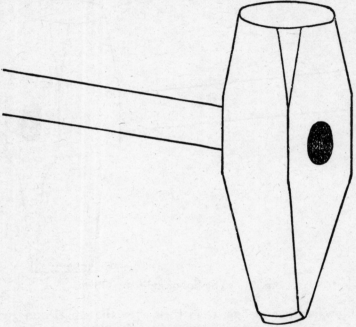

Fig. 22. The railroad chisel.

the rail than if the cutting edge was the full width of the body of the chisel.

Drills and Drilling.

Drills are in great demand, both in the machine and blacksmith shops, but where drilling by hand power is performed, is where good, fast and easy cutting drills are especially necessary. As in my first years at the trade, I had a great deal of practice drilling by hand, and with drills that were always breaking or being too soft, consequently putting the hole through often by main strength and energy, and other times with the drill squeaking by losing its cutting edge on the corners of the bit after drilling a few holes.

A great many blacksmiths prefer using machine-made drills rather than make a flat drill, because they say a flat drill will not do the work of a machine drill, and if they are asked the reason why they will say that the ones they buy in a hardware store are secret tempered or give some other reason, and often they will reply: "I don't know why." Now I wish to say to the reader that the author once made a flat drill that drilled 200 holes through a plate of hard metal ½ inch thick, and 40 holes through a plate of steel, at two grindings, without redressing. Will any machine-made drill do better? I will explain, although I mention one drill that did this amount of work. It was not tempered through guess work, neither did I strike the right temper by accident as some readers may think, as I will make any flat drill do the work of a machine-made drill, and I will go one better, as I will make a flat drill go through hard material which a machine-made drill will not penetrate. Reader, if you are an ordinary mechanic you can make a flat drill to do as well, if you will follow the instructions given in this chapter.

Making a Flat Drill.

Good tool steel of 75 points carbon will do for all ordinary drills, but for extremely hard drilling, steel of a higher carbon, say 90 points or 1 per cent, will be best. When making a flat drill, for example ½ inch in diameter, naturally the bit of the drill will be forged by making flat the end of a round piece of steel, but in forging the bit, do so if possible, by striking the bit edgewise as little as possible, and never try to forge the cutting edge of the bit, but cut off at right angles with a thin hot or splitting chisel on a piece of flat iron or copper. (This will keep the chisel from coming in contact with the hard face of the anvil, which would readily dull the cutting edge of the chisel.) Now strike the flat surface of the bit a couple of hard blows on each side while still at a low heat.

To have a fast cutting drill, the bit should be bent a little at right angles to form what is known as a lip, as in Figure 23, No. 1. Do this at a low heat (not

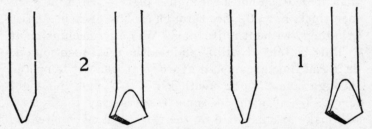

Fig. 23. Illustrating front and side views of flat drills, for hard and soft material.

exceeding a low cherry red), after the hammering has been done on the flat surface, by placing it a little over the edge of the anvil and then striking it with the hammer. This shape of a bit will cut soft metal much quicker and with greater ease than if left perfectly flat, and will equal any twist drill for fast cutting. Should the bit be a little too wide for the right size of the drill or the edges a little uneven, after being hammered and the lip formed, take it to the vise and file (or grind) on the cutting edge, but be careful to have both sides of the extreme point of the bit at the same angle, or the drill will not cut even and also cause all the strain to be placed on one cutting edge, which will have a tendency to break the drill if under heavy pressure.

Harden at a low even heat by dipping deep in the hardening bath, now polish, and draw the temper over the fire to a purple, making sure that the corners of the bit are the same temper as the point (as some blacksmiths allow the temper to run down, after not dipping deep enough in the bath to harden, and so the point of the drill is tempered while the corners are soft on account of the temper running down so rapidly). A drill that is made after these directions will do a great amount of work without regrinding if plenty of oil is kept on the cutting edge when drilling wrought iron or steel, cast iron requires no oil.

When drilling excessively hard material, such as tempered saw plate, chilled metal, etc., a perfectly flat iron drill, as No. 2, Figure 23, will be best, but it must not be drawn out so thin as a drill which is to drill soft material, and the cutting edge should not have so much bevel, draw the temper to a light straw. If this is found to be too soft, harden and draw no temper. Should this fail, heat the metal and lay a piece of brimstone on the exact spot which is to be drilled. But in heating tempered saw plate, casehardened plow mouldboards, etc., use as small a fire as possible so as not to draw the temper or hardness over more surface than just what is needed to drill the hole.

Thick bars of slate are very difficult to drill, especially when drilling such small holes as $\frac{1}{4}$ inch diameter. When drilling small holes in slate, make a **drill** somewhat after the same way as for drilling hard metal but temper to a dark blue, having very little clearance, and when drilling, clean the dust out of the hole often, and occasionally dipping the drill in water to keep it from becoming hot and drawing the temper.

Hand-Made Twist Drills.

It is possible to make twist drills by being forged by hand, but requires a little skill. But by following these instructions there will be no difficulty after making the first one, and drills made after the following method will be superior and outwear any twist drill that can be purchased in a hardware store.

To make a $\frac{1}{2}$-inch twist drill, take a piece of $\frac{1}{4}$-inch round steel, heat and flatten out to $\frac{1}{4}$-inch thickness and allowing it to widen out 3-16 wider than the steel, and the same length as the drill is to be. The twist is then put in by holding it at a bevel on the anvil while at a deep yellow heat, but be careful not to put in too short a twist; better put in a long twist, as it will get shorter as the drill is finished. Now take a low red heat the entire length of the twist, then with a very light hammer forge the twisted flutes edgeways, beginning at the back and following gradually towards the point. By this operation the flutes will widen on the outside while the centre will remain thin. By hammering lightly at a low heat the drill will come to the required shape and be perfectly round. If the hollow grooves are more or less uneven, file them out with a small round file, but be sure that the corners of the bit will cut the full size of the drill when ground.

Harden the full length of the twist, polish bright, and

draw the temper to a purple by drawing back and forth over the fire.

Making a Twist Reamer.

Of all the different kinds and shapes of reamers, a twist reamer properly made, takes the lead for fast and easy cutting. But like a hand-made twist drill requires

Fig. 24. Illustrating how reamer is forged, before being twisted.

a little skill. To make one (for example ½ inch in diameter), take a piece of ½ inch round steel. First forge the shank to go into the brace, then draw down round and tapered to about 3 inches in length, having it ½ inch in diameter at the large end and 5-16 inch at the small end. Then flatten out to ¼ inch thickness at the large end and ⅛ inch at the small end. To put in the twist commence at the large end which will be best

Fig. 25. The twisted reamer completed

done in a vise by the use of a wrench, but the twist in the small end can be done with the hammer by holding at a bevel on the anvil, then hammer as mentioned in making a twist drill until the reamer comes to the shape required. But bear in mind the twist is put in to the left, while to cut the reamer is turned to the right.

The cutting edge of the reamer is illustrated by "a, a, a, a," Figure 25. But back from the cutting edge to the groove shown at "b, b, b," when filing should be left a little smaller which will act as clearance thus allowing the extreme cutting edge to come in contact more readily with the work.

Harden as a twist drill and temper to a purple. Should the twist be put in a reamer as in a twist drill (which is put in to the right), the reamer will draw into the material too fast, and be liable to break it. Also bear in mind not to give too much clearance or the cutting edge will take hold of the material too readily and stick, for a ½-inch reamer 1-32 clearance will be plenty and for smaller sizes less will do. When filing on the cutting edge have a thin flat iron plate with different size of holes in it to form a gauge and so regulate the clearance. File out the hollow grooves very smoothly and evenly so as not to leave any thick spots in the cutting edge.

The Polishing Stick.

A polishing stick is made by taking a piece of pine or other soft wood, which should be round, about 1½ inches in diameter and 15 inches long. Wrap a coarse sheet of emery cloth or sandpaper around it, then drive in two or three carpet tacks to hold it in place. This will be much handier than a piece of loose sandpaper. Do not use sandpaper after it becomes worn smooth, as it will not brighten the steel satisfactory in order to see the correct temper.

CHAPTER IV.

How to draw out an axe—Mill picks—Butcher knives—Fine springs—Dirt picks.

How to Draw Out, Harden and Temper an Axe That Will Cleave a Hemlock Knot.

Almost every country blacksmith has had more or less axes to draw out (as the saying is), but there are very few who have a reputation of being able to dress an axe so that it will cleave a hemlock knot, without breaking or otherwise having the cutting edge bent over. And yet how many axes have been thrown into the scrap pile, after coming through the process the blacksmith gave it, by being cracked sometimes half way across the bit and occasionally a piece dropping right out, and every blacksmith who has had any experience repairing axes will know this statement to be true. I have known farmers and lumbermen who owned axes which were known to cleave a hemlock knot and still hold a good cutting edge, refuse to part with them when offered three times the amount that would purchase a new axe. I have had farmers offer me a dollar to dress one axe, while they could go to a hardware store and buy a new one for the same amount. "Why?" Because it would hold a keen edge and they could rely on it, and they were not afraid of breaking a piece out of the bit when chopping hard or frozen timber.

I remember drawing out an axe for a farmer, and

after grinding it I told him to try it on all the hemlock knots he could find. But being a little suspicious he asked me if I would go to the woods with him, which was but a short distance from the shop, and I consented to go. After reaching the woods I found some large knots and told him to go ahead and try his axe, but he hesitated and made no offer to do so, so I took the axe from him and chopped into a knot, and after striking a few good blows I handed him back the axe. The first thing he did was to examine the cutting edge as he was expecting to see a piece missing from it, but it was all there the same as it left the grindstone. Imagine his surprise. I saw him a few days after and the first thing he said was: "I have been using the axe ever since and cutting every hemlock knot I could find and the axe is as good as ever." Reader, I am not relating this experience to fill up an extra page in this book, but to impress on your mind the benefit of making good tools, and in this case especially an axe.

I will explain the way the ordinary blacksmith goes about to draw out an axe. He gets a low heat on the bit and commences to hammer it and entirely on the cutting edge. If the axe gets too wide he turns it up edgewise and drives it back straight again, sometimes loosening the steel, but works away until the axe is forged into a shape to suit him. Now he hardens and tempers it after this fashion, he places it in the fire, gets one corner at a white heat while the other is barely red, dips it in a tub of dirty water about an inch deep (but if it does not crack in the water it will shortly after). Then the temper runs down sometimes one color, other times another, while on the other hand the temper runs down so rapidly that the corners become soft and **only** the centre of the bit is tempered.

The causes for the axe cracking are—by uneven and overheating which is the main cause, while another cause will result from improper forging or hammering the steel unevenly when at a low heat, and so drawing or working the steel on the outside while the inside has not moved, and between the two there is a sort of tearing operation going on which will increase the tendency to crack when hardening.

Now let us dress an axe properly as it should be done. An axe that is rather thick is best for the process, as it gives us stock to work on. After preparing the fire, having the coal well charred and large enough to heat

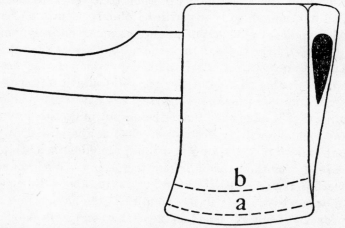

Fig. 26. Lumberman's chopping axe, dotted lines, a and b, indicate how to avoid strains in the steel when dressing.

the axe evenly the full width of the bit or cutting edge, heat evenly to a deep yellow; now commence to draw out by hammering, beginning at the cutting edge, going all the way across the bit, as indicated by dotted lines a, Figure 26, then turn over, going over this side the same way until the heat begins to get low and the edge is drawn as thin as is necessary for an ordinary chopping axe. Now we will take another heat, this time 1½ inches back and to a good yellow heat, but instead of going over the edge as before, go back 1¼ inches from the cutting edge, as indicated at **b**, Figure 26, and hammer both sides the same. Now the axe has widened out, but instead of turning it up edgewise and striking it with the hammer, cut it off with a chisel to the proper width. (A little narrower will be best as the axe is not finished yet, and the steel will come as wide as is necessary in the finishing stage and will also save filing.) Now heat the axe again, this time to a very low heat, just hot enough so that it will be plainly seen to be red when put in a dark place, and hammer entirely and evenly all over the flat surface and both sides the same for 1½ inches back after the same fashion as the first time, but this time hold the axe solid and level on the anvil and do not hammer the steel after it gets too cold, otherwise the tenacity that is being put in the steel with the hammer will be destroyed. The axe is now drawn out and after cooling off and filing the edges smooth, it is ready for the hardening and tempering. When hammering the axe in the finishing stage, about 15 good blows on each side will be enough. If the hammering is properly done at the right heat the steel will show a bright black gloss.

To harden the axe, heat the bit slowly and very evenly not less than 1¼ inches back, to a cherry red or just enough to harden, dip deep into the hardening bath not less than 1½ inches, raising and lowering so as to soften gradually, thus causing no strain in the steel. Polish the hardened part bright, but as there is not enough heat left in the axe to force the temper to run down, draw it by holding well over the fire, heating very slowly and moving back and forth so as to insure an even temper (but be careful and do not let the extreme thin cutting edge draw the temper first), until the whole bit of the axe for 1¼ inches back will show a light blue, then cool off and grind.

An axe that is forged, hammered, hardened and tempered after these directions, will be free from all cracks and will hold a very keen cutting edge, and if broken in use will be done by carelessness. When dressing an axe, the shape of the bit will depend upon the user, as some want a bit very rounding, others want the bit almost square.

When dressing double bitted axes, dress both ends before undertaking to harden and temper. When hardening and tempering be careful not to draw the temper in the first end when heating to harden the last end, but place the tempered end in water occasionally. A good chopping axe should be slightly thicker in the centre of the bit than at the corners, which will burst the chips more readily.

Mill Picks.

Mill picks are something that are very little understood by the average blacksmith, and I have known millers to send them 500 miles, in order that they might get them dressed and hardened properly. And as almost every blacksmith would like to know how, I will give the process. When making mill picks use steel of medium high carbon, say 90 points, as mill picks require harder steel than ordinary tools, as they have to cut very hard material, while the blows are very light.

There are different sizes of mill picks which depend

on the miller who is going to use them, ranging in weight from 2 to 4 pounds, but the medium-sized pick of about 3 pounds weight is most used. There are also different styles of mill picks, some have an eye punched in them for a handle, while others are made to fit in a socket. The main object, however, is to have a mill pick drawn out thin and hardened properly, in order to do a great amount of cutting and hold a good cutting edge, without being ground often or having the corners continually breaking off, which is dangerous to the miller's hands and eyes.

Now supposing some picks are to be dressed. The

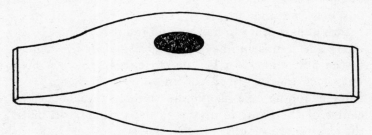

Fig. 27.   Correct shape of mill pick.

first thing to do is to draw the hardness by heating one end to a low red, before dressing the other end, otherwise there will be a tendency of the end held in the tongs to break off, when dressing the opposite end, unless the picks are very thick. Heat the end to be dressed or drawn out, to a deep yellow heat (so that the steel will be worked clear through, thus leaving no strains which would afterwards cause cracks), draw it out thin to 3-16 of an inch thick on the cutting edge, tapering to ⅜ of an inch thick, 1½ inch back, at the same time having the cutting edge barely as wide as the body of the mill pick. Now heat again to a low dark red, have a hammer weighing not less than 3 pounds, hold the pick perfectly solid and flat on the anvil without raising or lowering it, and strike 5 or 6 good blows on the flat side, making sure that the hammer will fall on the extreme cutting edge, then turn over and go over that side the same way. Do not strike every blow in the same place but go over the whole flat surface of both sides evenly and do not hammer the steel after it becomes black, but heat again to a very dull red and strike 3 or 4 more blows on each side evenly, and it is done. If the edges have spread out a little wide or uneven, do not attempt to strike one blow on the edge, but if necessary to have them straight, file or grind, although it makes no difference should the cutting edge of the mill pick be a little wide, as all the miller wants is to have them cut good. Now one end is dressed, go over the other end exactly the same way (before undertaking to harden), the cutting edge is filed on the same as for an ordinary cold chisel.

To harden a mill pick, heat evenly to a low cherry red or just hot enough to harden at least 1½ inches back from the cutting edge, then plunge into the hardening bath and cool until entirely cold. Then harden the opposite end after the same method, but be careful not to draw any temper in the end already hardened, and the pick is all ready for use. Do not draw any temper on mill picks as they will not be too hard if properly hardened. When hardening mill picks, harden one end at a time and never try to harden both ends at once. The right size of steel to use when making ordinary mill picks will be 1½ inches square, but for smaller sizes 1¼ inches square will do.

Butcher Knives.

There is a saying that has been going the rounds for a long time, which is: "Always use a file to make a butcher knife," but any blacksmith who believes in this saying or theory does very little thinking for himself. I have seen dozens of butchers' knives made from files, but what were they like? They were stiff, without any spring to them, were easily broken and would not hold a good cutting edge, on account of the teeth in the file, and to have a good butcher knife it must be just the opposite—it must hold a keen edge and instead of breaking, it must be very pliable, as whalebone. Although butcher knives are made almost entirely by machinery, machine or factory-made knives will not hold the keen cutting edge that a hand-made knife will when made from good steel and properly hardened and tempered. A great many machine-made butcher knives are made from poor and cheap steel, consequently they are sold cheap.

To make a butcher knife, use steel of about 75 point carbon. The proper size of steel to use for an ordinary size butcher knife will be ¾ by ⅛ inches. After the shape of the shank has been decided on, which may be flat, as a, Figure 28, forged or square as in Figure 29. Now, for example, we make a knife 8 inches long in the blade. The first thing to do (after cutting off the steel the right length) will be, take a chisel and cut off at right angles, as c in Figure 28, to form the point. Now crook the steel edgewise in a circular shape as Figure 28, the depth of the crook being about ½ an inch, then get a deep red heat half the length of the blade and draw to a thin edge, at the same time hammering evenly from both sides of the steel, beginning at **b**, Figure 28, on the inside of the crook, then heat the other half, and draw the edge until the point is reached. The crook that was put in the steel edgewise

Fig. 28. Piece of steel bent, drilled and pointed to make butcher knife.

has all come out and the knife is straight. Now flatten the point, making it thin and a little wider than the body of the knife. Never try to forge the point of the knife, but cut to shape with a chisel. Now take a very low heat the length of the blade of the knife, and hammer equally on both flat sides of the cutting edge to refine and pack the steel. Now heat again to the same low heat, but instead of hammering the cutting edge

Fig. 29. Illustrating carving knife, without handle, and forged with a square shank.

hammer in the centre of the blade, both sides evenly and the full length of the blade. Cool off and after filing the edges straight it is ready for hardening.

In hardening the knife have the fire a long shape, which will heat more of the knife at once. Now put the knife in the fire with the cutting edge downward, and move the knife back and forth so as to get a very even heat the full length of the blade, just hot enough to harden. Then plunge lengthwise with the cutting edge down into the hardening bath and cool entirely. Now, if the knife has been forged and hammered evenly the whole length of the blade, it will not come out of the hardening bath perfectly straight, but there will be a crook in it. For explanation we will say there is a crook in it. Now take a polishing stick and brighten one side of the knife. To draw the temper, place flat-

Fig. 30. Ordinary butcher's knife.

wise over the fire (do not place the knife directly on the hot coals, but instead, hold about 2 inches above), with the brightened side up. Heat very slowly, at the same time moving back and forth to insure a very even temper; draw to a pale blue, but before cooling place

Fig. 31. Straight pointed knife.

the knife on the anvil and take the crook out by striking it with a hammer, then cool, and it is ready for the grindstone. The reason that the crook was taken out so easily is, when the knife was being heated to harden only two-thirds of the width of the blade was heated enough to harden, consequently the back of the knife remained soft. This, with what heat was in the knife after the temper was drawn, allowed the knife to bend without springing back.

To have good success when making butcher knives, bear in mind to forge and hammer evenly on both flat sides, have a yellow heat when drawing the cutting edge, and a very low heat when hammering for the last time, and never attempt to strike the knife edgewise in the last hammering. Remember that the hardening of the knife must be carefully done so as to harden the full length of the blade without leaving any unhardened spots; also, as much pains must be taken when tempering so as to have it evenly drawn and so avoid having hard or soft places.

How to Make Gun, Revolver, Traps and All Fine Springs.

Springs perform a very important part in mechanical appliances, and especially guns, revolvers, and traps of all kinds, but to make a good spring the process must be thoroughly understood. When making springs, steel of about 60 points carbon is best, and never use steel for spring making that exceeds 75 point carbon, and bear in mind when making a spring to make it as wide and as thin as possible according to the work it is to do.

When making springs, as gun springs, in the first place forge the spring perfectly straight, but leave it a little less than the right width, also leave it a little thicker than the spring is to be when finished. Now heat the whole spring from end to end to a very low heat (say a dark dull red), and hammer with heavy blows evenly on both flat sides, being careful to keep the spring very straight without hammering the spring edgewise. As the hammering of the spring on the flat sides, when at a low heat, is to refine and pack the steel which is one of the greatest secrets in successful spring making. To bend the spring, heat it to a blood red (never exceeding a low cherry red), where the bend is to be, then bend to the proper shape. If at any time it is necessary to make a spring with a crook or offset in it edgewise, the crook must be put in the steel before the last hammering is done, so that the steel may be refined and packed by the hammer, thus increasing the tenacity of the spring, but do not hammer the steel below a certain heat, as the tenacity of the steel will be ruined when hammered too cold.

There are different ways of tempering springs, but only the simple and most successful methods are given. To harden, heat the entire spring in the blaze of the

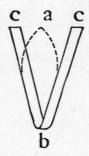

Fig. 32. Illustrating the elastic strength of a well made spring.

fire, very evenly to a cherry red or just enough to harden, then plunge into the hardening bath and cool off "dead cold." Then polish it bright. To temper, hold the spring edgewise 2 inches above the fire, and do not blow the fire, but heat very slowly, moving the spring back and forth and occasionally turning over to insure a very even temper, watching very closely until the spring has drawn to a very pale blue, almost grey. Do not cool off in water as other tools after the temper is drawn, but lay it down somewhere away from a draft of cold air allowing it to cool off slowly on its own accord. This is the author's favorite way of tempering a spring, and a spring as Figure 32, made after these directions, which for example, is 2 inches long and 1¼ inches between the ends c c, 1-16 thick and 5-16 wide, can be bent as indicated by dotted lines at a, until the ends meet, without fear of breaking or remaining bent, but instead will spring back exactly the same distance apart, as the ends were before being bent.

Still another good way of tempering a spring, after being hardened as formerly mentioned, and which will save polishing. Hold the spring over the fire, placing it in a dark place now and again, until it shows a very dark red, just visible to the eye. Then lay down and

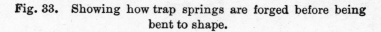

Fig. 33. Showing how trap springs are forged before being bent to shape.

allow it to cool off on its own accord, but in tempering a spring after this method have a dark place close to the fire.

When hardening or tempering a spring of any kind, hold it with the tongs at the extreme end, as the spring if held by the tongs in the centre, the place held directly between the tongs will have a great tendency to remain soft, which would spoil the spring and cause it to bend out of its proper shape when put into use. And bear in mind to harden and temper the whole spring evenly from end to end. A great many blacksmiths harden a spring in oil, but the error of this way is, the steel must be heated to a higher heat to harden in oil than water or brine, which is sure to decrease the elastic strength of the spring. In large factories where springs are made in great quantities, they are heated to harden, in hot lead or in a furnace, while the temper is drawn in boiling oil, tallow, or hot air. The degree of heat required to draw the temper is registered by a pyrometer connected with the vat or furnace, which insures a very fine and evenly tempered spring.

When making trap springs, or springs for a similar purpose after the fashion of Figure 32, have the spring at **b** a little wider than the rest of the spring and give it a short bend, which will increase the strength of the spring. Fine flat springs which have a long gradual bend when in use (especially for very light purposes), require no hardening or tempering as there will be enough spring to them, by simply hammering them equally on both sides when at a low heat.

Dirt Picks.

All tools of this description are used a great deal by railroad laborers, farmers and others, who work among gravel, hard ground, etc. Dirt picks are principally made from iron, while the ends are laid with steel which should be good ordinary tool steel of 75 points carbon. In dressing the ends, follow the instructions given in making or dressing a cold chisel, making sure to dip deep into the hardening bath, when hardening so as to give a foundation back from the cutting edge and keep the points or ends from bending and more especially if the pick is drawn out very thin, temper to a light blue.

For laying picks, I will give some instructions to insure good results and so produce a good weld. The steel for this purpose should be ¾-inch octagon or square, for ordinary picks, but if the picks are of a large size larger steel should be used. In welding of this kind always insert the steel into the iron, and don't make too long a weld, as a short one of reasonable length is best and much easier made. Always have plenty of stock when making a weld, so that the weld will need no upsetting in order to bring it to the right size.

Now draw down the end of the steel that is to be inserted in the iron in forming the weld, to a tapered or V-shape, and drawn to a thin square edge and having it at least ¼ of an inch wider than the weld is to be; cut a little piece out of the center after a V-shape and put some nicks in the beveled part of the steel with the corner of the chisel, making it very rough as No. 1, Figure 34. Cutting a piece out of the center of the

Fig. 34. Illustrating how ends are forged when laying dirt picks, in order to make a successful weld.

steel enables it to go well up into the iron. Take the iron part which is to form the weld, after heating, split it with a hot chisel as far back as required, open the ends and draw down to a thin edge as No. 2, Figure 34. Now take a good heat on the iron and insert the steel while at a low heat into the opening of the iron, driving the steel well up into the iron, then hammer the

ends of the iron well down over the steel. (The **nicks that** were put in the steel will enable it to stay in the **iron** much better during the welding heat, than if left perfectly smooth.) The pick is now ready for the welding heat. Have the fire clean with plenty of well charred coal, heat the place to make the weld to a cherry red, then place plenty of fine powdered borax on it, then heat to a welding heat as high as the steel will stand, but no more. Bring from the fire and strike the end against the side of the anvil or strike the end with the hammer, driving the steel well up into the

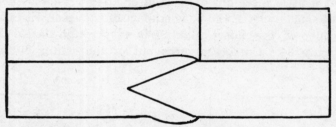

Fig. 35. How ends are placed together, showing plenty of stock to make the weld.

iron, then strike the first few blows (to make the weld) lightly and on the flat surfaces of the iron, then hammer hard and fast and from all sides alike until the weld is completed. If the weld is made after these directions, it will be perfectly smooth and solid and will show no trace or opening where the steel was inserted in the iron. The reason it will show no opening in the weld is because the steel was left wider than the iron and when being welded was driven back, filling the opening, that otherwise would have been there and so leaving the weld perfectly smooth and as solid as **one piece.**

Bear in mind that when making a weld after this description, have the iron well down over the steel as in Figure 35, and so leave no opening to allow dirt and other ingredients to get in when taking the welding heat, as steel will not weld successfully if any foreign matter gets in between the welding surfaces, no matter whether the weld consists of iron and steel or two pieces of steel. Also remember when taking the welding heat to keep plenty of fine powdered borax on it, so as not to allow the steel to become dry in the fire, and when making the weld strike the first three or four blows lightly, as the steel will unite more readily, than if the first few blows are struck very heavy.

When welding two pieces of steel and one piece is harder than the other, always if possible insert the hard piece into the soft piece, otherwise the steel is more apt to get overheated. And also bear in mind to take the welding heat of the harder piece, instead of the softer piece.

CHAPTER V.

Machinists' Tools—The use of asbestos and clay when hardening tools—Boilermakers' tools.

Machinists' Tools.

Connected with the machinist's trade is a wide range of fine and complicated tools. Among them being lathe and planer tools, milling cutters, taps, dies, reamers, etc., and the toolsmith in a large shop who can forge, harden and temper these tools satisfactory will have a great many friends among the machinists, but should the toolsmith be otherwise, there is nothing that tries the temper and patience so much as to have a number of machinists continually around the fire and each one making a complaint that the tool would not do this, and it would not do that, etc.

When making lathe and planer tools, there is no definite rule to go by "as to the shape," with the exception of a few standard or ordinary tools, as shown in Figures 36 to 46, as the machinist has so many jobs of a different nature that he requires a tool of a special shape to suit the work, which he explains to the toolsmith or gives him a drawing of it, and sometimes a pattern of it made from wood. Steel of 1 per cent or 100 point carbon is best for making lathe and planer tools, as these tools do their work by steady pressure. But nevertheless they must be properly forged, hardened and tempered, in order to stand the great strain that is continually bearing against them, and also to hold a good cutting edge in order to save time, and do work of a very exact and skillful nature, especially in the finishing stage.

Lathe tools of a flat surface, such as bent cutting off tools and side tools, must be forged to shape and the offset put in before the last hammering is done. When making cutting off tools and other tools of a similar nature, always give plenty of clearance as illustrated in Figure 38, which shows the cutting edge a a little

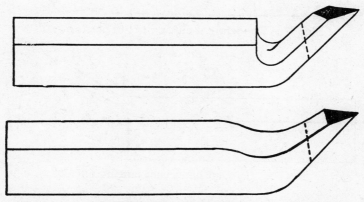

Fig. 36. Two ways of forging a diamond point lathe tool.

wider than **b**, also wider than **c**, as shown in front view of the same figure, and bear in mind when forging the tool leave it a little wider at the cutting edge than the exact width, as the hammering on the flat surface after the offset is put in, will bring the cutting edge to the right width. To more fully explain let us suppose the cutting edge of the tool is to be ¼ of an inch in width, but in order to refine and pack the steel, we will have the cutting edge 5-16 of an inch wide be**fore** undertaking to hammer on the flat surface. If the hammering does not flatten the steel to exactly the right size and so leave the cutting edge a trifle wider, so much the better, as the tool will have a little stock for grinding.

All lathe and planer tools for ordinary work should be quenched or hardened about one inch back and ac-

Fig. 37. The straight cutting off, or parting tool.

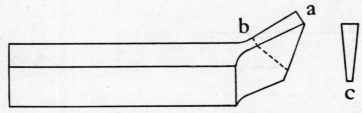

Fig. 38. The left hand bent cutting off tool.

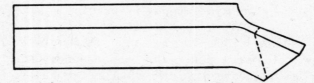

Fig. 39. The right hand bent cutting off tool.

cording to the cutting edge, as illustrated by dotted lines in Figures 36 to 45, and tempered to a dark straw or copper color. But should the tools have to cut very hard cast iron or other very hard material, harden and draw no temper. Should the tool still fail to cut the metal, use equal parts of powdered cyanide of potassium and prussiate of potash. To use this compound heat the tool as hot as if it was to be hardened, then place the heated, cutting edge of the tool into the powder, reheat again to a proper hardening heat and plunge into the hardening bath and cool off entirely.

Fig. 40. The spring tool.

When hardening large tools such as square corrugating tools, and which have very fine teeth in the cutting end, as indicated in Figure 46, make sure when heating that the center of the cutting face is as hot as the outside or corners. And never heat carelessly or too fast, so that the corners will be at a white heat,

Fig. 41. The left hand side tool.

while the center is barely red. But heat slowly and very evenly, until the whole face of the cutting end is just hot enough to harden, then plunge into the hardening bath and cool off entirely. Draw no temper.

Heat slowly and evenly until the whole face of the cutting end is just hot enough to harden, then plunge in the hardening bath and cool off. Ordinary corrugating tools require no temper drawn.

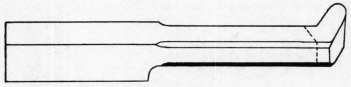

Fig. 42. The inside boring tool.

Air Hardening Steels for Lathe and Planer Tools.

Cast steel for lathe and planer tools is to a certain extent done away with in large shops by the use of air or self-hardening steels. A certain amount of this steel is manufactured in sizes so that no forging of the tool is required, and by the use of a patent toolholder (made especially for certain sizes of the steel) is su-

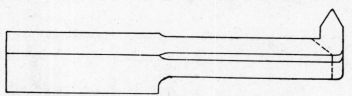

Fig. 43. The inside ordinary thread tool.

perior to a hand forged tool in some respects. But the toolsmith has an important work to do connected with air hardening steel, as he is called upon to cut it off in lengths and harden it, besides. As there is a limit to the sizes that are manufactured for immediate use the steel must be forged to the shape of the tool, while the forging and hardening of air hardening steel is somewhat of a different nature to cast tool steel. Air hardening steel is not as a rule used for anything but for roughing lathe and planer tools, it being too hard to make into any tool which is to do its work by the use

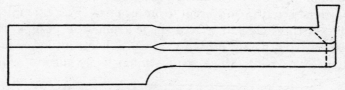

Fig. 44. The inside square thread tool.

of a hammer and also to make into any expensive tool such as milling cutters, as on account of its extreme hardness it cannot be machined or worked satisfactory. There are several makes and brands of air hardening steel, but some of the leading makes may be mentioned as, Sanderson's, Jessop's, Novo, Mushet and Blue Chip.

Fig. 45. The broad-nose or finishing tool.

When forging these brands into tools do not heat the steel above a bright yellow, especially Mushet or Sanderson's, but the steel must be heated evenly clear through the bar, and unless the tool is to be of a fine nature do the forging under the steam hammer, as the steel is too hard to forge with the hammer and sledge. The forging should be done as quickly as possible, while the heat is in the steel, and never attempt to bend or crook the steel when at a low heat, as it will be apt to crack or break in two, so be careful to have the steel at least at a deep red heat when it is to be bent.

When any new brand comes in the shop be careful to look at the directions on the bar, as some brands are hardened a little different from others. For example, the cutting edge of a tool made from Blue Chips is hardened by being heated to a white heat until it commences to melt, when small bubbles or blisters will

Fig. 46. The corrugating tool.

form on the steel, then it is placed in a blast of cold air. To harden Mushet steel heat to a deep yellow and cool off in a blast of cold air, Sanderson's and Jessop's after the same process. Novo steel is hardened by heating it to a white heat, then cooled off in a blast of cold air or may be quenched in oil.

When hardening steel with a blast of cold air, have the pipe or nozzle (which conveys the air) as close to the fire as possible, and when cooling off the tool, have some arrangement to hold the point or cutting edge directly in front of the blast. If this is not done the tool is apt to become turned to one side by the force of the blast. Also bear in mind if the steel is heated to a melting heat, as Blue Chip, be careful not to put the blast on too strong at first or it will blow the point off the tool. Instead put on the blast light and gradual until the steel begins to cool a little. Then turn on the blast to its full capacity and keep it on the tool until it is perfectly cold.

A great deal is to be learnt in working air hardening steel in order to get the best results. But for the beginner who has had no experience, follow up the directions given by the manufacturer. Should the directions fail to give good results do some experimenting. For illustration Novo brand of steel will give good results if heated to a white heat, then plunged into oil or boiling water, while Blue Chip will sometimes give better results if (after being heated to a melting point and cooled off in a blast of cold air) it is heated to a very low black red heat and allowed to cool of its own accord. Yet there are other brands of steel which are hardened by heating to a light yellow heat, then placed in a cool place and allowed to cool slowly of its own accord.

Never allow Mushet steel to come in contact with water or it will crack, although no visible heat can be seen in the steel. Figure 47 illustrates a heavy roughing tool, made from air hardening steel for turning locomotive tires, car axles, etc. Dotted line at a indicates how far back from the cutting edge, as b, the steel is to be heated when hardening. All roughing tools are made after the same shape. Bear in mind when forging air hardening steel, never try to forge it below a low cherry red and if at any time it is necessary to cut a bar into lengths, do not try to cut it cold, as it must be heated.

How to Anneal Air Hardening Steel.

Sometimes, although seldom the toolsmith is called upon to anneal air hardening steel, so that it may be turned or planed. This class of steel is very difficult to anneal on account of its extreme hardness. It must not be packed directly in slacked lime or ashes as cast tool steel, as it will cool off too quickly. To anneal air hardening steel, use an air tight heavy iron box, place the steel inside the box, then heat altogether to a deep red heat, then pack the box deep into slacked lime or ashes. If an iron box cannot be obtained, use a heavy iron pipe or band large enough to accommodate the

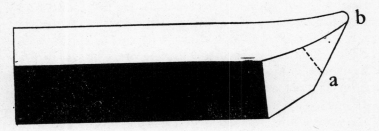

Fig. 47. Heavy roughing tool.

steel, without having the steel project out through the end. Also have two heavy flat iron plates large enough to cover the end of the pipe. Place the steel in the pipe and heat the pipe, likewise the plates, as formerly mentioned, then pack the pipe into the lime in an upright position, having one of the plates directly under the pipe, the other on top, which will keep the steel from coming in contact with the lime. By following this method the steel will keep hot a very long time, and thus give good results. To anneal a piece of 1-inch square air hardening steel will require from 12 to 15 hours' time.

Milling Cutters.

These tools are very valuable to the machinist's trade, and the making of them requires great skill, valuable time and good steel, which makes them very expensive tools. If they are not properly forged and hardened there is a great loss.

When forging a blank for a milling cutter (steel of about 90 point carbon is best) be sure and leave the blank a little larger every way than the exact size of the milling cutter when finished, as a machinist always prefers a little extra stock, which enables him to machine the tool with greater ease and less caution than if forged to the exact size of the tool. When forging a milling cutter heat the steel evenly to a good yellow, but do not heat too fast or the outside or corners will be at a white heat while the inside is barely red, and by unevenly heating will cause strains which will produce a tendency to crack when hardening the milling cutter.

If the milling cutter is large forge it to shape under the steam hammer being careful to forge it evenly from all sides alike and so work the steel clear through, but as it becomes finished reduce the heat. Tools such as milling cutters which are either round or almost of an equal size, cannot be refined, as the steel will remain at its natural state, consequently it is not desirable to finish at quite so low a heat as those of a flat surface. But if the milling cutter is very thin and flat then the steel may be refined by using a flatter on the flat surface while at a low heat.

After forging the milling cutter it must be annealed, which operation must not be neglected, as the whole blank must be heated again slow and evenly to a blood red (never exceeding a cherry red), then packed deep into slacked lime, allowing it to remain there until cold. Have the annealing box large enough so that it will contain plenty of lime to keep the heated steel away from the air. If there have been any slight strains left in the steel by forging the annealing will take them out and make the steel soft so that it can be worked with ease by the machinist.

The hardening of a milling cutter after it has been machined to shape, is the process that must not be overlooked, as in the hardening the toolsmith must either succeed or spoil the tool. The milling cutter must be hardened properly without cracking, so that when put into use it will do a great amount of cutting by holding a good cutting edge, and so it is necessary that the toolsmith use great care when heating for hardening. If there is no heating furnace in the shop, and the milling cutter is to be heated for hardening in the coal or open fire, too much care cannot be used.

When heating to harden in the open fire have the coal well charred and the fire plenty large enough. The top of the fire should be perfectly flat and the whole surface a perfectly and very evenly heated mass, place the cutter (if after the shape as b, Figure 48) flatway on the top or surface of the fire, now heat very slowly and evenly and turning it over occasionally until it is heated to a very even low cherry red, or just enough to harden, then plunge it into the hardening bath edgeways from a vertical position, allowing it to remain deep in the bath until quite cold, then dry off and polish bright.

To temper the milling cutter, take two round bars of iron about two feet long and just large enough to go through the hole in the center of the milling cutter, put one end of each iron into the fire and heat to a white heat to about two inches back from the end or according to the thickness of the tool. Now take one of the irons (leaving the other in the fire) and place the heated end directly in the center of the milling cutter, holding there until the temper in the teeth has drawn to a dark straw color. If the iron cools to a low heat before the temper is drawn to the exact color use the other iron which was left in the fire to finish drawing the temper, then cool off. Should the milling cutter have very heavy teeth or if it is to cut very hard metal, it will not be necessary to draw any temper.

If the milling cutter be to the shape and size, say 4 inches long and 2½ inches in diameter (or larger sizes) as a, Figure 48, a good way to heat for hardening in the open fire is to have a heavy iron pipe about 6 inches long and plenty large enough for the cutter to go inside. A pipe about 1 inch wider than the diameter of the milling cutter will be about right. Place the pipe in the fire and build the coal on top of it, but do not build the fire over either end of the pipe. Instead leave the ends of the pipe open and do not allow the coal to get inside the pipe. By this method an opening is left clear through the fire. Heat the pipe to a bright cherry red the whole length, insert the milling cutter and heat it to the necessary heat to harden, then plunge into the hardening bath from a vertical position. But when heating to harden a milling cutter which has very fine teeth after this method, be very careful not to bruise the fine cutting edges of the teeth against the pipe. It is better to have some arrangement to hold the milling

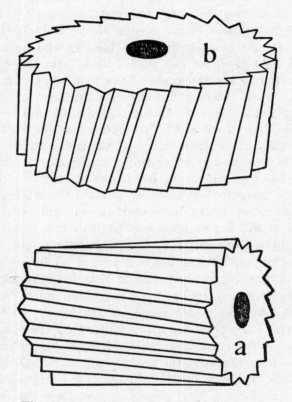

Fig. 48. Plain or ordinary milling cutters.

cutter up from the pipe. An iron bar placed through the center of the cutter with a bearing under each end will keep the cutter from coming in contact with the pipe.

The Use of Asbestos and Clay, When Hardening Milling Cutters and Other Tools.

Very often a milling cutter is made with a thread through the center of the tool which must be kept soft, while the outside or teeth are hardened, and the way this process is accomplished is by the use of asbestos, which is packed well into the inside or thread, but make sure that the outside ends of the thread are well padded over without allowing the asbestos to come in contact with the cutting edges of the teeth of the milling cutter. The asbestos is kept in place while hardening by the use of fine pliable wire, wrapped around the tool. After the hardening has been done and the asbestos taken out from the inside the thread will be quite soft. The reason the thread has remained soft while the teeth are hardened is because the water could not come in contact with the thread when being quenched, on account of the presence of the asbestos.

I have saved a great many delicate and expensive milling cutters from cracking when hardening, by the use of asbestos. Take for example an angle end milling cutter. That is made with a thin or delicate part extended from the main body of the tool. Now, although the tool may be very evenly heated and properly hardened, it is still very liable to crack, and in some cases the thin or extended part will crack off in a solid ring. To stop a milling cutter of this kind from cracking, fill the hollow in the end, as in Figure 49, with asbestos, being careful not to cover any of the cutting edges of the teeth and hold the asbestos in place by the use of fine wire, while hardening. The reason the tool will not crack is, when quenching to harden only the teeth side comes in contact with the water which hardens, while the other side is kept soft as the asbestos keeps the water from coming in contact with the hot steel.

Another example: Take a piece of steel 3 inches long and 1 inch thick, now an inch on each end is to be hardened, while the remaining inch in the center is to be kept soft, and to accomplish this process, wrap the center well with asbestos, keeping it in place by winding some fine wire around it, or instead of using asbestos, wrap the steel around with clay, keeping it in place by the use of a thin piece of sheet metal wound

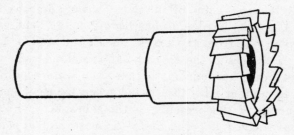

Fig. 49. The angle end milling cutter.

around it, then heat to harden and the results will be as formerly explained.

Hardening Hollow Tools.

When hardening milling cutters as a, Figure 48, spring threading dies or any similar tool, always quench them from a vertical or upright position, which will allow the steam and water to come up through the tool and cause the steel to be hardened more evenly. Should the tool be quenched from a horizontal position it will be impossible for the steam to escape, and which will keep the water from coming in contact with the hot steel. Thus when the water is held back by the steam there is a tendency for soft spots in the tool.

As a rule steel workers never pay any attention to the steam when hardening, which is a great mistake, as many tools are partially if not altogether spoiled (more especially if the tools are of a delicate nature) by the great amount of steam which rises as soon as the hot steel comes in contact with the water. Delicate or fine tools of a hollow nature will sometimes warp or even crack, caused by steam and improper methods of quenching.

When hardening a spring threading die, it is not necessary to harden the whole tool, but just far enough back from the thread to allow the temper to be drawn with safety, as indicated by dotted line in Figure 50.

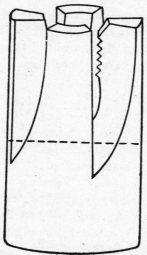

Fig. 50. The spring threading die.

To draw the temper, after being hardened as just mentioned, hold the end or thread part of the tool above the fire and draw the temper very slowly and evenly (by keeping the tool turned around) to a dark straw color.

The Hardening and Tempering of Hob Taps, Stay Bolt Taps and Similar Tools.

In forging, annealing and hardening of long slender tools, such as hob taps, stay bolt taps, etc., too much care cannot be exercised, although as a rule these tools do not have to be forged, as the steel is generally obtained from the manufacturer the right size to allow it to be machined into the tool. However, the steel should be well and evenly annealed, should it come direct from the manufacturer or should it be forged by the toolsmith.

In annealing as well as hardening long slender tools they must be carefully handled when the tool is heated the whole length of itself or it will warp easily, also pack the tool very carefully when annealing so that it will have an equal bearing.

When hardening get the tool to a very even heat, enough to harden the whole length of the thread, and when quenching dip deep in the center of the hardening bath from a perfectly upright position, allowing it to remain in the bath until perfectly cold. Bear in mind that when quenching a long slender tool, any variation from a perfectly upright position will have a tendency to warp the tool. When hardening do not harden the shank, as all that is required to be hardened is the cutting teeth or thread.

To draw the temper, polish bright the grooves in **the**

thread from end to end, have a couple of heavy wrought iron pipes or bands. Heat both in the fire to almost a white heat, then remove one from the fire and put it in a convenient place, place the tap in the heated pipe and draw the tap back and forth to insure a very even temper from end to end. If one pipe is not sufficient to draw the temper, replace with the hot pipe that is in the fire, then cool off the tool. For all kinds of taps draw the temper to a dark straw color. Be sure when drawing the temper not to use too small a pipe, or the extreme fine points of the thread will draw too quickly. For all ordinary taps use a pipe about five inches long and 3 inches inside diameter, while the thickness of the pipe should not be less than $\frac{1}{2}$ an inch or it will cool off too quickly.

Sometimes when hardening a long slender tool only a certain part of it is to be hardened. For example, supposing we have a long slender tool 18 inches long and 1 inch thick. Now 6 inches in the center is to be hardened, while 6 inches at each end is to be kept soft. In a case of this kind take an iron pipe 7 inches long and 2 inches inside diameter, build the pipe into the fire a little above the surface of the forge, and heat the pipe evenly all around and from end to end. Now place the tool through the pipe, having the part which is to be hardened directly in the heated pipe, while the ends which are to be left unhardened will project from each end of the pipe, which will prevent them from becoming hot enough to harden. To keep the tool from warping or bending (while being heated) place something under each end close to the pipe to form a bearing and also to keep the tool in the center of the pipe. If the tool is to be hardened to a very exact length wrap with asbestos at the ends of the part which is to be hardened, as the asbestos will prevent the steel from becoming hardened while being quenched. To illustrate this more clearly, **a a**, Figure 51, represents the asbestos, **b** the part of the tool which is to be hardened and **c c** the unhardened ends.

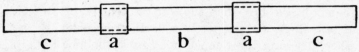

Fig. 51. Illustrating how certain parts of tools are hardened.

Heating Furnaces.

In large shops and factories where tools are made in great quantities, furnaces are used to heat steel, but principally for hardening purposes, and in a great many respects a furnace is superior to an open fire, as steel can be heated in a furnace very evenly and with less danger of it becoming overheated than if heated in an open fire. Tools of a long slender shape, such as stay bolt taps, and tools of a very wide and flat surface such as milling cutters, are best heated in a furnace. Different kinds of fuel are used for heating furnaces. The principal ones used are gas and oil. Gas, however, is preferred, as the furnace can be very readily regulated in order to heat the steel to any degree of temperature. There are many different makes of furnaces and different methods of operating them.

Heated Lead for Hardening Purposes.

The lead bath is extensively used for heating steel when hardening, and has many advantages that a **fur-**

nace does not possess, as in the lead bath certain parts of tools may be heated in order to harden with ease, and the temper drawn in many ways which could not be accomplished with a furnace.

When heating steel in lead, be sure to use a chemically pure lead, containing as little sulphur as possible. Sometimes when heating in lead there is danger of it sticking to the tool when hardening, but to overcome this difficulty make use of the following compound: Take a pound of powdered cyanide and dissolve it in a gallon of boiling water, afterwards allowing it to cool. Now dip the articles to be hardened in the liquid, remove them and allow to dry before placing them in the lead. The liquid when allowed to dry on the tools will form a moisture on the tools when in the heated lead and prevent the lead from sticking. If lead is allowed to stick to the tool while hardening it will cause soft spots where the lead remains. When heating tools with fine projections have a fine brush and clean off the tool should any lead happen to stick to it.

To obtain the best results when heating in lead, keep the lead stirred up, as it will always naturally be the hottest at the bottom.

A lead bath is preferred to an ordinary heating furnace, as steel heated in lead will not raise a scale; also if the lead is heated to the proper degree it will be impossible to overheat the steel, consequently the steel will be very evenly heated. Should there be a great many small tools to harden at once, place them in a heavy wire basket or sieve and lower into the lead; after being heated they may be all quenched together.

Boilermakers' Tools.

Boilermakers' tools, although not so many in number or of so complicated a nature as machinists' tools, should be well understood in order to forge, harden and temper them successfully. The principal tools used by boilermakers are beading tools, punches and dies, rivet snaps, flue expanders, drifts, calking tools and chisels.

The Beading Tool.

This is a tool which the average toolsmith does not understand, especially the hardening and tempering. Beading tools are principally made from 7/8-inch octagon steel of 75 point carbon. After forging, then filing or grinding to the proper shape as in Figure 52,

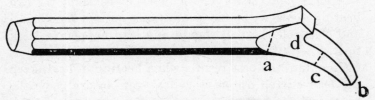

Fig. 52. Correct shape of beading tool.

harden as indicated, between the dotted lines, then temper to a light blue. A great many tool-dressers harden the whole end of the tool from **a** to **b**, and the results in a great many such cases are that the tool breaks off at **c**. Now I wish to say to the average toolsmith that the point acts only as a guide for the tool and it should be kept soft, as only the hollow in the tool at **d** does the work and only the part that requires to be hardened and tempered.

When a great many of these tools have to be hardened and tempered at one time, a good way to do this work is by having a very small fire, just large enough to heat the part to be hardened. The fire should not be any wider than 1½ inches across the surface, and the way to build a fire for this purpose is, after the fire is nicely started build up with wet coal, with the exception of a very small place exactly in the center, which should be kept well filled up with fine crushed coke or charred coal, keep the blast blowing gently as the fire is being built up. By this method a fire may be made very small.

When only a few beading tools are to be hardened at one time and it is necessary to heat them in the ordinary fire, keep the point of the tool from **b** to **c** well cooled off in water. Do not allow it to become a white heat before cooling it off. Cool it often, not allowing it to become hotter than a low dull red. In this manner the point will be kept soft, while the part between **a** and **c** is heated to a cherry red so as to harden. When quenching dip in the hardening bath down as far as **a**. To draw the temper let it run down very slowly, or what is better, draw the temper over the fire. To obtain the best results from tempering the whole space between the dotted lines should show a light blue.

Punches and Dies.

When hardening punches and dies (for perforating boiler plate or iron) and when a great many of these tools are to be hardened at one time, there is nothing better to heat them in than heated lead, but as most of shops are without this convenience likewise a heating furnace, we must resort to the open fire. However,

speed will be acquired and time saved when **hardening
and** tempering in large numbers if the following method is put into practice: Have the fire large and flat and well heated across the surface. Now place a plate of iron or boiler plate which should be about 9 inches in width, directly on top of the fire, bank the coal up around the edges of the plate to about 3 inches high and in the form of a circle, leaving the circle or open space on the plate about 5 or 6 inches across. Now place as many punches or dies on the hot plate inside the circle with the cutting edge up as is convenient. Then place another plate over them, allowing it to rest on the bank of coal which will form a furnace. Turn on the blast slowly and as the punches or dies become hot enough to harden, quench them and replace with others until all are hardened.

To draw the temper, polish bright and place them on a hot place with the cutting edge upwards and allow to remain until the necessary color appears on the cutting edge. Then cool off. For a punch allow the temper to draw to a light blue. For a die a dark straw will be good. When heating small punches and dies for hardening as Figure 53, heat the whole tool and quench from a vertical position.

Flue Expanders.

When hardening long flue expanders pins, as Figure 54, follow the instructions as given in hardening hob taps, stay bolt taps, etc., but harden the whole length of the tool from **a** to **b**, while the temper must be drawn to a dark blue if the right kind of steel is used, which should be about 75 points carbon, likewise all small parts of flue expanders should be tempered to a dark blue.

All small parts of flue expanders may be heated to harden, by placing them on a hot plate of iron. The plate of iron being placed directly on the surface of the fire and heated to a light yellow, a slight hollow in

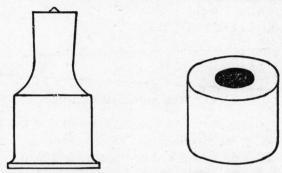

Fig. 53. Punch and die for perforating boiler plate, iron, etc.

the iron plate will be best, which will keep the small parts together, and also keep them from rolling off the plate, but bear in mind to occasionally turn the tools over in order that they will be heated evenly. To draw the temper can be done somewhat after the same

Fig. 54. The expander pin.

method as heating to harden, but do not have the plate of iron nearly so hot, as a low dull red heat in the plate will be sufficient to draw the temper. However, keep the tools well turned over in order that they will be evenly tempered.

Drifts, Rivet Snaps, Calking Tools and Chisels.

When making a boilermaker's drift it must not be hardened or tempered or it will break easily. It should be forged to a long gradual taper very round and smooth. For a half inch drift take a piece of ⅞ round or octagon steel, draw it down to ¼ of an inch at the small end and tapering back to ⅞ at the large end, while the length is 5½ inches.

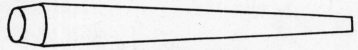

Fig. 55. The boilermaker's drift.

Rivet snaps, which are used for rounding the heads of rivets when riveting, are made from ⅞ octagon steel. The end as marked **a**, Figure 56, is made by upsetting the steel, while the hollow in the end illustrated by dotted lines is generally made by a machinist by turning it to shape in the lathe. Harden and temper it after the fashion of a cold chisel, draw the temper evenly to a light blue by gradually and slowly turning the tool around above the fire.

Calking tools are made somewhat like a cold chisel, with the exception that the bevel of a calking tool is on one side only, the other side being perfectly straight with the body of the tool.

Boilermakers' chisels will be found mentioned in another part of this book.

Hardening Shear Blades.

When hardening blades to shear boiler plate or iron, heat the cutting edge only (say for 1 inch back) by drawing back and forth through the fire, making sure to get a very even heat from end to end, then quench in the hardening bath from a vertical position and cool off entirely. When tempering, polish bright and draw the temper to a purple by moving back and forth over the fire or on a hot plate of iron, being careful to get a very even temper, then cool off. Sometimes, although seldom a shear blade is made with two cutting edges instead of one, so in a case of this kind, heat the whole tool evenly from end to end when hardening, afterwards tempering as already mentioned.

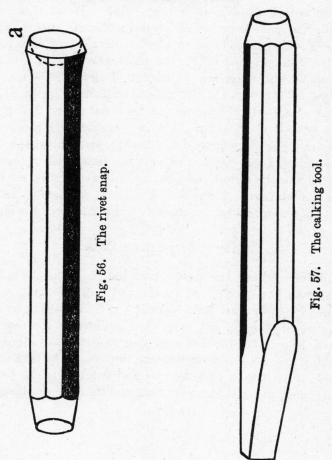

Fig. 56. The rivet snap.

Fig. 57. The calking tool.

CHAPTER VI.

Woodworkers' and carpenters' tools.

One of the greatest secrets of success connected with woodworkers' tools, is to be able to make them so that they will hold a keen cutting edge. Woodworkers' tools are made now almost entirely of steel, with a few exceptions, such as a carpenter's chisel, which is made from iron and laid with steel. But when laying a carpenter's chisel or any similar tool don't select any piece of scrap steel that might chance to be handy, such as an old file because it is of a flat shape, but get the best steel of about 75 points carbon. If there is no flat piece of good steel in the shop near the shape of the chisel, forge a piece of octagon down flat to near the shape the chisel is to be, as it is always better to do a little hard work; spend a little extra time and produce a good tool, in order to get a good reputation.

Laying a Carpenter's Chisel.

When making the weld upset the iron and steel in order to have plenty of stock, and when forging the scarf for a separate weld have it a little rounding as a, Figure 58, but do not leave any hollows in the scarf, as b, Figure 58, for dirt and slag to get into. Before taking the welding heat prepare the fire, by having the coal well charred and the gas and sulphur taken out. Now heat the iron to a raw heat and take a good borax heat on the steel, after drawing from the fire (before placing together to make the weld), strike the ends on the anvil to knock off all foreign substance. When hammering to make the weld strike the first two or three blows lightly, then with heavy blows, making it a point to finish the weld by striking on the flat side when at a low heat. Also bear in mind that when hammering to make the weld strike the first few blows di-

The correct method.

rectly in the center, which will weld the center first and force all slag and foreign matter out. Otherwise if the ends are welded first the slag will be forced into the center and cannot get out, which would keep the parts from uniting or welding.

After welding to finish the chisel follow the direc-

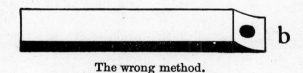

The wrong method.

Fig. 58. Illustrating methods of scarfing steel for separate welding.

tions as mentioned in making a cold chisel, after hardening draw the temper to a dark blue. These directions will apply to all wood chisels, plane bits, etc.

When making or repairing any tool with a beveled

cutting edge, such as a framing chisel, Figure 59, forge the bevel to shape as indicated by a, which will save a great amount of filing or grinding, and the tool will be just as good if it is hammered when at a low heat on the flat and beveled surfaces only. Harden and temper all ordinary woodworkers' chisels, at least 1½ inches back from the cutting edge.

Fig. 59. The framing chisel.

The Screw Driver.

A good screw driver is a tool which is prized by almost every mechanic who uses one, but more especially by woodworkers, and there are very few blacksmiths or tool dressers capable of making one to give perfect satisfaction, as the screw driver will generally break or twist when coming in contact with hard screw driving.

When making one select good steel of 75 points carbon. After it is forged to the correct shape for an ordinary screw driver as Figure 60, strike it a few good

Fig. 60. Correct shape of ordinary screw driver without handle.

blows on each flat side of the screw driving edge while at a low heat. Then harden and allow the temper to draw to a grey, cool off and you will have a screw driver that will give unlimited satisfaction.

How to Make a Draw Knife.

A draw knife is a very handy tool for woodworkers, and is also used a great deal in general blacksmith shops. Although draw knives are made almost exclusively in large tool factories, they are as a rule inferior to one made properly by hand and from good steel. Draw knives vary in shape and size according to the work they are to do, but also a great deal depends on the fancy of the woodworker who is to use it.

To make a draw knife for ordinary use and which will answer mostly all purposes take a piece of steel of 75 points carbon, 9 inches long, ⅜ thick and ⅞ wide. To forge, first of all heat and bend the steel edgewise in a circular shape as in Figure 61, then draw down the cutting edge from the inside of the bend to within an inch of each end, indicated by **a a**, which will bring the knife back straight again. Now fuller in as illustrated by dotted lines **a, a**, then draw down the ends to form the shanks as **b, b**. Now draw the bevel on the back of the knife which will crook the knife edgewise again, but it does not signify, as the knife will come straight again before it is finished. Now weld on pieces to the shanks to form the handles, but instead of welding on straight, then bending, take a piece of small round iron ¼ thick, upset the end of the iron and weld as illustrated at **c, c**. After welding, bend the handles up a little at right angles in order to allow the draw knife to be finished (as it must be done in the open fire). To finish the knife, heat the whole length of the cutting edge of the knife to a low dull red, then hammer the whole flat surface of the cutting edge from end to end, then turn over the knife and hammer the other side the same way. Now take another very low heat as before, but instead of hammering directly on the cutting edge, hammer in the center of the blade and on both sides the same. After hammering the knife has come straight again, but if it is not exactly straight and instead there is a little fullness in the cutting edge it will be better than if perfectly straight. When making an ordinary draw knife have the under side very flat.

To harden the draw knife, have the fire wide or in a long shape, with the coal well charred and loose. Place the knife in the fire with the cutting edge downwards, turn the blast on very slow and move the knife back and forth through the fire to insure a very even heat the full length of the cutting edge. Then quench by plunging into the hardening bath with the cutting edge downwards, and cool off entirely. If the knife was well forged and hammered evenly on both sides the knife will come out of the hardening bath perfectly straight, but otherwise there will be a crook in it flatwise. However, supposing there is a crook in it, polish one side bright and draw the temper to a light blue by holding the knife flatwise over the fire and moving back and forth in order to get a very even temper, "but before cooling off" place the draw knife on the anvil and take the crook out by striking it with the hammer. The reason the crook is taken out so easy is because only two-thirds of the knife was hardened, the back of the knife remaining unhardened, which along with the heat being in the knife, allows the hardened and tempered part to bend without springing back, now cool off and the knife is ready to grind.

After the tempering is done, to bend the handles to shape, heat to a dull red, being careful not to draw the temper in the main body of the knife. This will be overcome if the body of the knife is placed in water or by keeping it cool with a wet rag or sponge.

Piece of steel bent, to make draw knife.

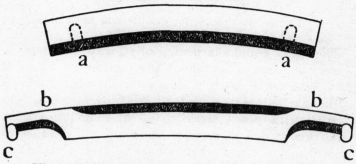

Illustrating, cutting edge drawn, back beveled and shanks forged.

Draw knife completed with handles welded on.

Fig. 61. Showing how to make a draw knife.

To hold the handles on a draw knife, generally the shanks are riveted at **d, d**, but what is better, cut a thread on the ends of the shanks and fasten the handles on by the use of a small nut and washer.

CHAPTER VII.

Stonecutters' tools for granite and marble.

The making and sharpening of stonecutters' tools is a very important branch of the toolsmith's art, and first class tool dressers of stonecutters' tools are scarce. Although the greatest obstacle to be overcome in connection with stonecutters' tools, after having a thorough knowledge as regards the nature of steel, is to know the shape and what temper to give. But otherwise the steel is worked practically the same as in making a cold chisel, or any other flat tool which is to have a cutting edge.

There are many branches to the stonecutter's trade, which varies according to the class of stone or rock the stonecutter is to cut or work on, as some classes of stone are much harder than others, consequently there is a great difference in the ways of cutting different classes of stone, and also as much difference in the shape and temper of the tools. The most common classes of stone which are cut and used for building purposes are granite, marble, limestone and sandstone.

Granite Cutters' Tools.

The tools principally used to cut granite are points, chisels, bush hammers, mash hammers, granite hammers, bull sets and bull chisels.

In Fig. 62 is illustrated the correct shape of a point and chisel, which are drawn down very thin at the extreme cutting edge but with a very short taper. When dressing chisels, keep a very coarse file on hand to straighten the cutting edge after hammering. Points and chisels are made as a rule from ⅞ octagon steel, but points will be better if made from ⅞ square, as a point made from square steel will be held much easier to the stone, and will not twist round in the hand so easy as if made from octagon steel.

Fig. 62. Correct shape of granite point and chisel.

When hardening points and chisels, it is not necessary to heat them hot enough to harden any farther back than ½ inch, but be careful when heating so as not to get the extreme thin cutting edge too hot. When quenching to harden dip deep into the hardening bath to at least one inch, so that the temper will draw very slowly to a dark straw, which is good for all ordinary granite tools. But if the granite is **exceptionally** hard, draw no temper on the chisels.

The Granite Drill.

Fig. 63 illustrates front and side views of a granite hand drill, which are used almost entirely in quarrying granite, and is made from ¾ octagon steel. The shank of the drill is drawn down to ½ inch round and 4½ inches long. The bit is forged to a square or diamond flat point, being left 1-16 thick on the cutting edge and ⅝ wide, which gives the drill a good clearance and enables it to cut much better and faster, and will not get fast in the hole. Harden and temper as a point or chisel already mentioned. A drill forged,

Fig. 63. The granite hand drill, illustrating flat and side views of the bit.

hammered, hardened and tempered properly, will drill two holes 3½ inches deep. I have had them put in as many as four holes 3½ inches deep, but a drill improperly hardened or overheated will not put in one hole, besides tiring the driller much quicker.

Bull Sets and Bull Chisels.

Bull sets and chisels are dressed and hardened as a stone hammer, with the exception that only one end is dressed, the other end being a chisel head which is left unhardened for the sledge to strike upon. A bull set is made the same shape as the face end of a stone hammer. The bull chisel resembles the flat or tapered end. Stone hammers will be found fully explained and illustrated in another chapter of this book.

The Granite Bush Hammer.

The cutting part of a granite bush hammer is comprised of thin flat blades, which are held in place by

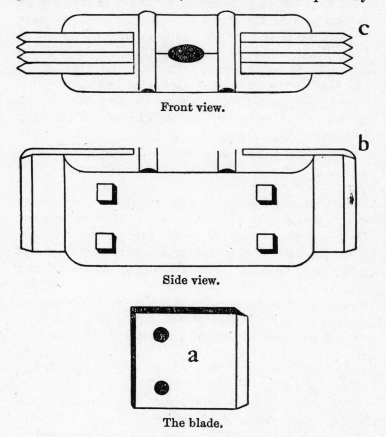

Front view.

Side view.

The blade.

Fig. 64. The granite bush hammer.

two bolts going through the blades and also through the hammer. The blades are taken out to sharpen, as shown at **a**, Fig. 64, the cutting edge of the blade, which consists of a very short level from each side, is forged or hammered to sharpen, while the corners are left somewhat rounding, which keeps them from breaking off, as shown in figure at **b**. When sharpening be careful to keep the blades perfectly straight and have the cutting edge beveled equally from both sides, also bear in mind that the blades should all be exactly the same length after being sharpened and replaced in the hammer, as illustrated by dotted line at **c**, otherwise should the blades be uneven the hammer will not do good work, while the blades will have a tendency to break at the cutting edge.

In order to make the corners rounding when sharpening, strike a blow on the corners edgewise before hammering on the beveled surface. Then the steel will be refined and the cutting edge sharpened by striking every blow on the beveled surface. After hammering, file the edges straight with a coarse file.

As the cutting edge of the blade is very thin and wide, be very careful when heating to harden by having a very even heat. Heat hot enough to harden about ¾ of an inch back from the cutting edge, but when quenching dip the blade to at least 1 inch deep, then polish one side bright and draw the temper over the fire, moving the blade sideways back and forth to insure a very even temper, draw to a light straw, then cool off.

Some bush hammers differ slightly from others, owing to the number of blades they contain. The one illustrated in Fig. 64 is considered a coarse cutting hammer.

The Granite Hammer.

A granite hammer resembles an ordinary stone hammer, with the exception that both ends of a granite hammer is drawn down very tapering to an extreme sharp cutting edge, as Fig. 65. After hardening, temper to a light straw.

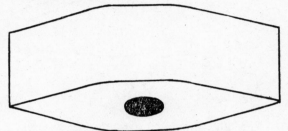

Fig. 65. The granite hammer.

Granite Cutters' Mash Hammer.

When making a granite cutter's mash hammer, use good steel of 75 point carbon, and 1¾ inches square. When forging, do not put in a large oval eye as other hammers, but instead punch a small round eye not exceeding ¾ of an inch in diameter, which is about the right size. The ends of a mash hammer are beveled down from the eye on the top and sides, the bottom side is left perfectly flat. The face at each end is also on a bevel, which should be dressed a little full in the center and the corners left very rounding. Harden both ends as mentioned in making hammers in another chapter of this book, but after proper hardening it is not necessary to draw any temper as there are no sharp corners to break off. Fig. 66 illustrates the correct shape of a mash hammer.

The Granite Tool Sharpener's Hammer and Anvil Stake.

When sharpening granite tools, the toolsmith should use a bevel face hammer and anvil stake, as Figures

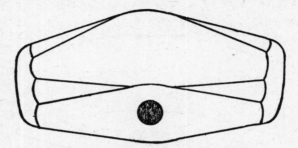

Fig. 66. Granite cutters' mash hammer.

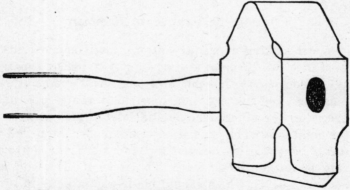

Fig. 67. The granite tool sharpeners' hammer.

67 and 68. These tools are a great advantage over an ordinary flat face hammer and the plain anvil, and after a little practice in use the work will be performed much easier, quicker and neater.

The stake sits in the square hole of the anvil and should be kept perfectly firm, by driving a flat key through the shank illustrated at b, the shank being made long enough to project through the anvil 1½ inches. Harden the face surface of the stake all over but not very deep. Be careful when heating to harden

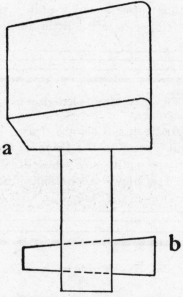

Fig. 68. The anvil stake.

so as to get a very even heat in order to harden the center of the face, and do not heat too quick or the corners will become overheated. The beveled front side of the stake at **a** will be found much better than if made perfectly straight for cutting off the ends of broken tools and uneven edges.

To make a granite tool sharpener's hammer, use

steel 1¾ inches square. To harden and temper follow the directions given in "Hardening and tempering a hammer" in another chapter.

Marble Cutters' Tools.

Tools for cutting marble are in a great many ways the same as those used to cut granite, with the exception that the temper is left a little harder, say, a very

Fig. 69. Marble lettering tool.

light straw. Lettering tools are drawn out very thin, as Fig. 69, but owing to the thinness be very careful when heating, and make sure when dressing to always strike the last few blows on the flat side only, when at

Fig. 70. Marble tooth chisel.

a very low heat. Tooth or ordinary plain chisels are not drawn out so thin as lettering tools. When making or dressing tooth chisels, the teeth are filed in after the hammering is done.

CHAPTER VIII.

Stone cutters' tools continued, for limestone and sandstone—Stone lathe and planer tools.

Limestone Cutters' Tools.

As limestone is of a soft nature, the tools used to cut it are of a very different shape and temper to those used for hard stone such as granite or marble. In cutting soft stone of any description wooden mallets are used, consequently the heads of the tools are made broad and a little rounding, so as not to cut or bruise the mallet, as illustrated in Fig. 71. This class of tools is known as "mallet headed tools."

The tools used principally for cutting limestone are plain and tooth chisels, points, pitching tools, hand and ball drills, tooth axes and bush hammers. When making any of these tools use steel of 75 points carbon.

Plain and Tooth Chisels and Points.

Chisels and points are made chiefly from ¾ octagon steel. When making chisels, if the chisel is to be narrow forge the head first, but should the chisel necessarily have to be made wide, the chisel end should be forged first, as when forging wide chisels the steel will have to be upset (according to the width of the chisel) before drawing down to a thin cutting edge. The teeth are put in tooth chisels with a punch machine, after the hammering is done on the flat surface, and then filed to an extreme sharp point.

When hardening chisels be careful to get a very even heat all along the cutting edge and dip in the hardening bath not less than one inch; polish bright and draw the temper to a very pale blue, which is the right temper for all limestone tools.

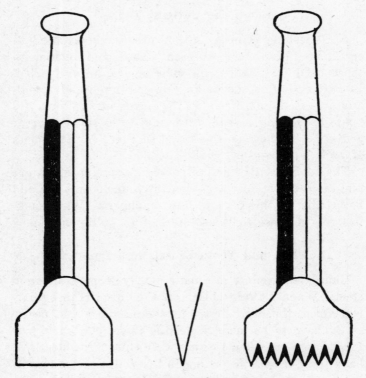

Fig. 71. Plain and tooth chisels for limestone.

Points are drawn down to almost a square point. When hardening dip deep into the bath, and draw the temper so that it may be seen at least ⅝ of an inch back from the cutting edge, otherwise should the temper run down very fast so that only the extreme point becomes tempered, the point will be easily bent and broken.

Fig. 72. The limestone point.

Pitching Tool.

A pitching tool is made from one inch octagon steel, but to form the cutting edge the steel must be upset a great deal. The cutting edge of a pitching tool for all soft stone, should be slightly beveled as indicated by a, Fig. 73. The cutting edge of a granite pitching tool will be best if left perfectly straight.

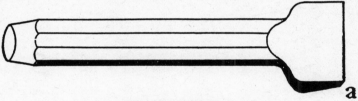

Fig. 73. The pitching tool.

Hand and Ball Drills.

Hand drills for limestone are made on the same principle as hand drills for granite (see Fig. 63) with the exception that the bit on a drill for limestone is drawn out much thinner and to a sharp cutting edge, the cutting edge being made slightly rounding.

Ball drills are made by taking a bar of iron about 5 feet long and 1¼ inches in diameter, then welding a piece of ¾ octagon cast steel at each end, afterwards drawing out as a hand drill, thus having a drill at each end. But bear in mind that the drill on one end should be ¾, the other ⅝. A ball drill is used for drilling by raising it up, then allowing it to drop down from a vertical position, the larger drill on one end being to start the hole, and the smaller one at the opposite end being used to finish. Thus the large drill gives clearance to the smaller drill, which prevents it from sticking in the hole. After hardening, draw the temper to a pale blue.

The Tooth Axe.

The tooth axe is a tool used a great deal by limestone cutters, and which has the cutting teeth at each end as illustrated in Fig. 74. To make a tooth axe

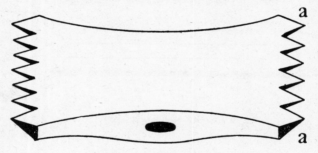

Fig. 74. Side and end views of tooth axe.

take a piece of steel 2½ by 1½ and 5½ inches long. After the hole is punched, when drawing down the ends allow the steel to gradually widen out towards the cutting edge. Draw down both ends to a flat and sharp cutting edge before undertaking to cut in the teeth.

Before cutting in the teeth, have a block of wood or some other convenience to rest one end of the tooth axe on while cutting the teeth in the opposite end and so keep the cutting edges of the tooth axe from becoming bent or bruised up. As some tooth axes have more teeth in one end than the other, measure off and nick in a little with a sharp cold chisel where the teeth have to be cut in. Now take a deep red heat and cut in with a thin splitting chisel as deep as is necessary at the nicks as measured off, afterwards cutting the pieces out at right and left angles and the shape of the teeth will be formed. But bear in mind when cutting the teeth make sure that the outside teeth, as a a, Fig. 74, are a little the largest or heaviest, which will add strength and keep them from breaking off (owing to the extra hard usage) when in use.

Finish at a low heat by placing a flatter on the flat surface and let the helper strike it a few good blows, which will straighten the teeth and add strength and tenacity to the steel, allow the tooth axe to cool off, then file off all roughness between the teeth by using a three-cornered file, also file the teeth to an extreme sharp point, making sure that all the teeth are in a straight line and exactly the same length.

When hardening, heat all the teeth evenly, being careful not to overheat the corner or outside teeth, dip to quench in the bath to about 1¼ inches deep, and draw the temper over the fire, moving back and forth in a sideward motion to insure a very even pale blue temper.

The Limestone Bush Hammer.

To make a bush hammer that is to give good satisfaction, requires great skill and care in forging, hardening and tempering, but if these instructions are followed closely good results will follow. To make one, use steel 2 inches square and 7 inches long; after punching the hole, which should be small and almost

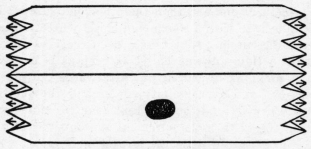

Fig. 75. The limestone bush hammer.

round (say ¾ in. diameter), now before cutting in the teeth have the end of the hammer perfectly square and flat, and to form the outside of the teeth the hammer should be beveled equally on all four sides as shown in Fig. 76. But remember that after the end of the hammer is forged, and the teeth are to be cut in, never heat above a bright cherry red, otherwise, if a very high heat is used when cutting in the teeth, the teeth will not hold a good cutting edge, as there is no way to refine the steel.

Bush hammers differ somewhat, by having different numbers of teeth, which range from 16 teeth or 4 cuts to 144 teeth or 12 cuts. The one illustrated represents a 4-cut hammer. To put in the teeth first measure off into squares according to the number of teeth required, and nick in with a sharp edged cold chisel; now heat the hammer to a cherry red and proceed to cut in the teeth by cutting in at the nicks with a thick edge chisel, which will form the point of the teeth. Heat again; this time take a very thin edged splitting chisel and cut in to the depth required, which will vary according to the size of the teeth, but for a 4-cut hammer, cut in to ⅝ of an inch deep. The first end of the hammer may have the teeth cut in by letting

Fig. 76. Illustrating how bush hammer is forged before cutting in the teeth.

the opposite end rest on the anvil, but to cut the teeth in the other end the hammer will have to be placed on a block of wood or some other convenience, which will not interfere with the teeth.

After the teeth have been cut in they will have to be filed to a sharp square point by using a flat featheredge file, thus, ◆ . A file of this description will go much deeper into the cuts than an ordinary three-cornered file. When filing be careful to keep all the teeth the same length. In large up-to-date tool shops the teeth are put in by the use of a planer, but in small or ordinary shops the teeth must be put in by the use of a chisel and file.

When hardening, heat to a very even heat just enough to harden, being careful that the corners do not become overheated by heating too fast, and also watch the extreme points of the teeth for fear they become overheated. Although only the teeth are necessary to be hardened, it is always best to have plenty of heat back of the teeth in the main body of the hammer (which should not be hot enough to harden, and so should not exceed a low blood red), which is necessary to drive down the temper, as the temper on the inside teeth cannot be drawn over the fire unless extreme great care is exercised. After hardening by dipping vertically in the bath to the depth of almost one inch, moving up and down a little at the same time in order to soften gradually, then polish up bright as well as possible and allow the temper to run down in the teeth to a very light blue.

When tempering bush hammers, as a rule, the teeth in the centre are apt to draw the temper first, while the corner teeth are yet hard. To overcome this difficulty, take a small piece of wet rag or sponge fastened to a piece of wire, commonly known as a swab, and as soon as the temper in the centre teeth draws to a light blue, place the wet swab directly on the tempered part. This will hold the temper in check in the centre teeth until the outside teeth draw to the proper temper.

Harden and temper both ends of the hammer the

same, but when hardening and tempering the last end be careful not to draw the temper in the opposite end (which is already hardened and tempered), which can be overcome by dipping in water occasionally.

Sandstone Cutters' Tools.

Sandstone, although of a soft nature, differs somewhat from limestone, it being of a very sandy composition, while limestone is in the nature of very hard clay. Tools properly made for limestone will remain sharp a long time, but tools for sandstone wear away very fast. The tools used to cut sandstone are practically the same as those for cutting limestone, the points and plain chisels being the same, but tooth chisels differ somewhat as the teeth are left flat, as shown in Fig. 77, while tooth chisels for limestone are filed to a

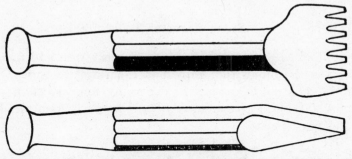

Fig. 77. Showing sandstone tooth chisel and splitting tool.

sharp point. The teeth in sandstone tooth chisels are put in with a punch machine after the last hammering is done.

When dressing plain or tooth chisels for sandstone, be careful to leave them a little rounding in the cutting edge, as these tools when cutting sandstone are naturally inclined to become hollowing in the centre, and more especially in wide chisels. When tempering sandstone tools let the temper draw to a light blue, but when tempering very wide chisels be careful and do not allow the temper to run out at the centre while the corners are at a straw color, as all wide tools of an irregular shape (especially when not being hardened very far back from the cutting edge by not being dipped very deep in the hardening bath) have a tendency to draw the temper in the centre first, so have a small swab which consists of a small piece of wet rag or sponge attached to the end of a piece of wire, and when the temper in the centre of the chisel reaches a light blue, press the wet swab directly against it, which will hold the temper in check until the corners are drawn to the proper color.

Stone Carvers' Tools.

All stone carvers' tools are made very thin and fine, but as the blows that fall upon them are exceedingly light it is not necessary to draw any temper, as they will stand and give good results, if properly hammered and hardened.

Polishing Board for Stone Cutters' Tools.

When tempering stone cutters' tools and a great many at one time, have a short piece of board nailed or fastened in a convenient place close to the forge, and have some fine clean sand to put on the board. The board, if beveled a little upwards, will hold the sand on it much better than if fastened perfectly level. After hardening, the tools are brightened while in the hand or tongs by simply rubbing on the sand across the board.

How to Forge Mallet Head Tools.

To forge or make mallet head tools, they should be first upset a little, afterwards being fullered in, then drawn out. Then swage to shape by using tools as shown in Fig. 78.

Fig. 78. Top and bottom swage for making mallet head tools.

Punching Teeth in Stone Cutters' Tools.

There are different kinds of punch machines used in performing this work, which are operated by an upright, side or foot lever, and all will do good work as a rule if the punch and die are made properly. The correct shape of these tools are shown in Fig. 79, which illustrates both die and punch made with a bevel. This style of punch and die will punch the teeth very easy and the punch will not break so easy and will keep the chisel straight when punching in the teeth. The opening in the die at a should be a little larger on the under side, which will act as clearance. The cutting edge of the punch should fit the opening very closely, it being a little thinner back from the cutting edge in order that it will move up and down through the opening "in the die" with ease. The beveled cut-

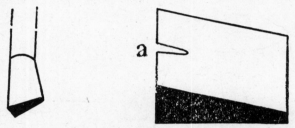

Fig. 79. Punch and die for teeth punching machines.

ting edge of the punch allows it to punch the teeth by a gradual cut, and which works much easier than if it was made with a perfectly straight cutting edge.

The teeth will punch best when the steel is just a little warm. When punching be careful to hold the chisel perfectly level on the beveled surface of the die, otherwise the chisel will bend. In cases where the teeth bend after punching, do not undertake to straighten again when cold, as that will take out all the toughness and tenacity from the steel (that was put in the steel when hammering at a low heat) and the chisel or the teeth will break very easy when in use, but when straightening after being punched heat the chisel to a low heat.

Harden the punch as any ordinary flat tool and draw the temper to a dark blue. To harden the die heat the whole tool evenly enough to harden, then cool off entirely, afterwards drawing the temper over the fire or on a piece of heated flat iron. Draw the temper to a dark straw.

Lathe and Planer Tools for Cutting Soft Stone.

In large and up-to-date stone yards, a great amount of stonecutting is done by machinery, and which brings into use "stone lathe and planer tools." There are a great many tools of this description used, and almost every one is of a different shape, varying according to the work that is to be done. Some of these tools are made from heavy steel, as finishing tools shown in Fig. 80, which are made from 1 by 4-inch flat steel,

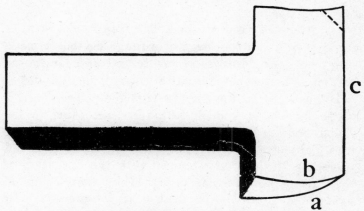

Fig. 80. Finishing tool for stone planer.

while the cutting edge is about 6 inches wide. **To** make a finishing tool as illustrated, the steel will have to be upset, then drawn or flattened out until it is the proper size. Heat to a good yellow heat when forging. The cutting edge is drawn down on a bevel principally from one side, as illustrated by **a** in figure, leaving the other side, **b**, perfectly flat. After the cutting edge is drawn out (as just mentioned), then hammered at a low heat to refine the steel, the cutting edge at **c** should be bent up a little from the flat side, which will enable the tool to be cut much better. Roughing tools illustrated in Fig. 81 are made some-

Fig. 81. Roughing tool for stone planer.

what after the same method, but from much smaller steel; also the cutting edge of a roughing tool is made rounding. Moulding tools are forged as near to the correct shape as possible, afterwards being ground.

To harden, heat the cutting edge about three-fourths of an inch back, enough to harden, according to the shape, but dip deep in the bath and draw the temper to a light straw color. For finishing tools, it is not necessary to draw any temper with the exception of the extreme corners, as indicated by dotted line of figure, which should be drawn to a light straw by holding over the fire.

Dressing Tools with the Cutting Edge Beveled from One Side Only.

When dressing tools of this description a great difficulty is to be contended with, such as hammering or forging equally from both sides. Most blacksmiths and tool dressers have a great tendency to hammer on the beveled side only, which is a great mistake. Take, for example, stone planer tools as illustrated in Figs. 80 and 81, if they are dressed by hammering entirely on the beveled side, they will be sure to crack when hardening the second or third time they are dressed, and the tendency to crack will be increased if forged entirely at a low heat, so bear in mind to have a good yellow heat when commencing to forge and finish at a low heat, but forge or hammer equally from both sides.

CHAPTER IX.

The stonemason's nammer—Miners' tools.

The Stonemason's Hammer.

A stone hammer is the mason's favorite tool, and which almost every general blacksmith is called upon to dress, but there are very few who understand the process thoroughly so as to give good satisfaction, and knowing that almost every general blacksmith would like to be capable of doing this work successfully, I will give these instructions, and if followed closely the trouble with stone hammers will be over. Have a large fire with the coal well charred (as it is poor policy trying to heat a stone hammer in a small fire); now place the face end of the hammer in the fire and heat slowly to an even yellow heat clear through the hammer for $1\frac{1}{2}$ inches back. Be careful not to heat too fast so that the corners will be at a white heat while the centre is almost black. Uneven heating of this description will cause strains in the steel and have a tendency to crack when hardening.

When dressing a stone hammer always finish at a low dull red heat. After the face end is dressed go over the flat end in the same way, with the exception that the last blows should fall on the flat sides only when at a very low heat. To form the cutting edge on the flat end of a stone hammer it can be forged to shape with the hand hammer or cut to shape with a thin splitting chisel.

TOOLSMITH AND STEELWORKER 147

Now that both ends are dressed, return the face end to the fire; heat carefully and slowly until the whole face is a cherry red. Never allow the corners to get above a cherry red when heating to harden and should the corners reach the necessary heat to harden before the centre of the face, stop blowing the fire, which will check the heat in the corners and allow the centre to come up to the required heat. To harden, plunge the whole hammer deep into the hardening bath and cool off entirely, at the same time keeping the water agitated so that it will keep cool and act on the hot steel much quicker. Heat and harden the other end in the same way, but be careful not to draw any temper in the end already hardened, which can be overcome by occasionally dipping the hardened end in water while heating the other end, for all ordinary stone hammers never draw any temper.

As almost every blacksmith or tooldresser who has

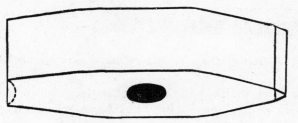

Fig. 82. The stone hammer.

dressed stone hammers has had more or less experience by having them crack, bear this in mind, that it is not the water which causes them to crack, as a great many may suppose, "but instead by the fire," for if a stone hammer becomes overheated in the fire, especially when heating to harden, there will be cracks in it, and if they don't show up the first time after quenching, they will the next, and this information likewise applies to other tools. Sometimes when working on soft stone, some masons prefer a stone hammer made with a hollow face, as indicated by dotted lines in Fig. 82.

Miners' Tools.

As there is a great amount of mining going on throughout the country, first-class sharpeners of miners' tools are in great demand, and likewise receive high wages. The principal tools used by miners are drills, which may be divided into two classes, as, the hand drill and the cross or machine drill.

To make a good fast cutting hand drill, notice the shape of the drill in Fig. 83. I have had drills of this

Fig. 83. Correct shape of miners' hand drill for hard rock.

description drill a hole 19½ inches deep, through hard granite at one sharpening, and for a whole day at a time I have had them average 14 inches to each sharpening, but yet how many drill sharpeners are there that can make a drill average 6 inches in hard granite without resharpening? The real secret of success in order to make a good fast cutting hand drill is this: The drill must be properly hardened but not very far back from the cutting edge, and have plenty of clearance on the corners of the bit. The cutting edge is

TOOLSMITH AND STEELWORKER   149

formed by drawing down to a very short taper, and should be gradually rounding towards the extreme cutting edge (instead of being perfectly straight). As shown in side view of Fig. 83, the cutting edge should also be just a little rounding.

A great many drill sharpeners do not give their drills enough clearance on the corners and at the same time making the drill very rounding in the bit as Fig. 84. Others again make the drills too thick at the

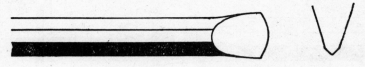

Fig. 84.   Incorrect shape of hand drill.

cutting edge, as shown in side view of the same figure. Miners call this shape a bull bit because it will neither cut or break on account of its thickness. This shape of a drill is made by driving back the steel with the hammer when sharpening. Ordinary hand drills for hard rock such as granite, requires no temper to be drawn after hardening.

The Cross or Machine Drill.

The machine drill which does its work by the aid of steam, is entirely different from a hand drill, the bit or cutting edge being in the shape of a cross, as shown in Fig. 85. To make a machine drill, the end is first upset according to the size of the drill to be made, then split in from four sides with a thin hot chisel, afterwards fullering in with square fullers, which will leave the end as Fig. 86. The bevel of the cutting edge is then cut on with a hot chisel.

When forging or dressing the bit of a machine drill, make sure that the centre of the bit is exactly in the centre of the drill, so as not to have one part of the bit longer than another, and have the bit perfectly square and all the cutting edges exactly the same length; also have a good clearance on the corners. Although machine drills are used chiefly in mines they

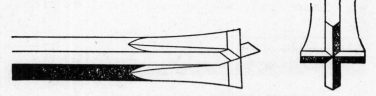

Fig. 85. Illustrating side and end views of a machine or cross drill.

Fig. 86. Showing the shape of a machine drill bit before cutting bevel to form cutting edge.

are also used in stone quarries, but when making a machine drill for soft rock such as limestone, the bit should be made thinner and should also have a longer bevel to form the cutting edge, than a drill which is to drill hard rock.

When hardening a machine drill, heat the whole bit evenly but not exceeding ¾ of an inch back from the cutting edge, then plunge into the hardening bath and cool off entirely. Drills to cut hard rock require no temper drawn, but a machine drill to drill soft rock should be drawn to a dark blue.

The Breaking of Drills when Drilling and the Cause.

The main cause for drills breaking is due to over and uneven heating and also by having too long a heat when hardening, but a great many times the drill will break and although the broken fracture shows a very solid, fine and close grain in the steel representing a piece of glass (a sure sign that the steel was properly hardened), there must be some other cause, and in cases of this kind I used to blame the drill runner for carelessness, when at the same time I was the one to blame for not making the drill the correct shape by not giving the drill enough clearance on the corners.

It should be remembered that if a drill has not enough clearance and binds in the hole it is very easily broken, although it may be properly hardened. Another cause for drills breaking is, when drilling rock that has seams or cracks running through it, but this is a natural cause and cannot as a rule be remedied, although it may be partly overcome when drilling with a hand drill, by making the drill perfectly straight in the bit. When drilling holes 12 or 15 feet deep, the hole is always started with a large drill, but as the hole is drilled deeper the drill will have to be made smaller and longer. A good rule to go by so as to regulate the size of the drill for deep drilling, is to make the bit or cutting edge ⅛ of an inch smaller to every 2 feet in length of the drill. To further explain, supposing a hole is to be drilled 16 feet in depth, the first drill or starter will be 2¼ wide in the bit, while the last one to finish the hole will be 1¼ inches wide in the bit. Generally a gauge is kept on hand for the purpose of regulating the size of the drill, which consists of a piece of thin flat iron having the different sizes cut in it.

The Rock-Cutting Reamer.

Sometimes after drilling, it is found necessary (by the man who does the blasting) to break the rock a certain way, but as satisfactory results cannot be accomplished by the ordinary round drill hole, the hole is cut into an oval shape and in a direction which is most likely to cause the necessary results. This work is known as "reaming the hole."

The reamer is made as a rule from octagon steel, by upsetting according to the size of the drill hole to be reamed, and forged to an oval shape. The long way of the oval cutting face should be a little rounding, while the ends as illustrated, **a a**, Fig. 87, should

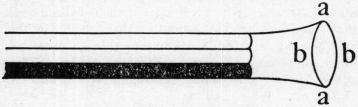

Fig. 87. The rock cutting reamer.

be forged to a sharp point, and the cutting edges, as **b b**, should be perfectly square and sharp. The reamer should be well tapered back from all sides of the cutting face to give a clearance.

To harden, heat the cutting face enough to harden about ¾ of an inch back, then plunge into the bath and cool off "dead cold" and draw no temper. Unless the reamer is to cut soft rock, then draw the temper to a dark blue.

Well Drills.

There is a great amount of well drilling going on throughout the country, consequently the blacksmith is called upon to dress the drills occasionally. The size and shape of the drill depends upon the size of the hole and the hardness of the rock to be drilled. A drill for hard rock is made thick and heavy, and is hardened without drawing any temper. But a drill for soft rock is made thin, and after hardening the temper is drawn to a dark blue. When making wide and thin well drills be careful in heating to forge, but more especially when hardening.

CHAPTER X.

Horseshoers' Tools—How to dress a vise—Sharpening plow shares.

Horseshoers' Tools.

After considering the number of horseshoers there are, it is safe to say that not one in a hundred can make or dress their own tools as they should be done. I have been in a great many horseshoeing shops, where I have seen men working by main strength and energy simply for the want of good tools.

The majority of horseshoers buy their tools in a hardware store, using them until they become dull, then they are thrown away, because the horseshoer does not know how to fix them. But as every horseshoer likes to have good sharp tools, I will give the following instructions on different tools which if put into practice, it will be no longer necessary to work with dull tools.

How to Make and Dress a Pair of Pincers.

To make a pair of pincers, take a piece of ⅞-inch square steel which should be 75 points carbon, forge and bend to shape as illustrated in Fig. 88, leaving the jaw the full width of the steel and tapering towards the cutting edge, but be sure and leave the jaw heavy and strong as indicated by a in Figure 88. After the jaw is bent to shape, strike the flat surface of the cutting edge a few good blows by placing the jaw on the narrow or flat end of the anvil. The handles may be

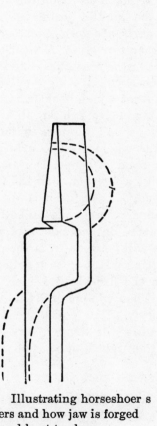
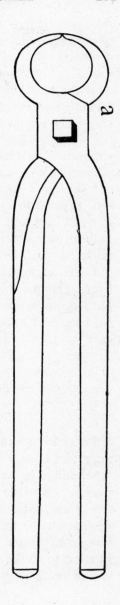

Fig. 88. Illustrating horseshoer's pinchers and how jaw is forged and bent to shape.

drawn out round or half round as suits the man who is to use them, but will be much easier on the hand and be a pleasure to use if the handles are made ⅝-inch half round. After both parts are forged fit together and drill the hole for the bolt or rivet. A steel bolt is preferred to a rivet, as then the pincers can be taken apart very easy whenever necessary.

To harden, heat at least 1 inch back from the cutting edge and draw the temper to a light blue. To dress pincers, hoof cutters or nail nippers, without taking out the rivet, heat the whole jaws to a bright cherry red, then close in or bend to shape. Now have a very low red heat and strike a few blows on the flat surface only of each cutting edge, although it is always best to refine the steel by hammering equally on both flat sides, pincers can be hammered on one side only after being bent to shape, which will be done from the outside. To harden, have the cutting edges close together, so as to heat both at once in the fire about 1 inch back from each cutting edge. Then plunge into the hardening bath and cool off entirely. Now polish the inside of the cutting edges, then draw the temper very slowly and evenly over the fire to a light blue.

Making a Clinch Cutter.

To make a clinch cutter, take a piece of good ordinary tool steel 1 inch wide, ⅜ inch thick and 5 inches long, fuller in as shown in Fig. 89, then draw out the part to form the handhold. Now forge the edge for the clinch cutter, making it a point to finish by hammering on the flat sides of the cutting edge while at a low heat. Draw out the punch almost square, bearing in mind to strike the last two or three blows on the flat side when at a low heat.

Harden the clinch cutter by heating the cutting edge about ⅝ of an inch back, then quench in the bath as indicated by dotted lines in Fig. 90, then polish one side and draw the temper over the fire to a light blue. Harden and temper the punch after the same method as the clinch cutter. Bear in mind that when hardening the

Fig. 89. Illustrating piece of steel fullered in to make a clinch cutter.

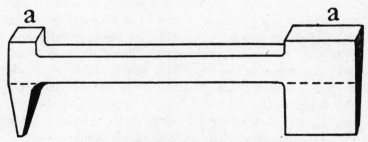

Fig. 90. The clinch cutter completed.

clinch cutter or punch, keep the back as indicated by a a, Fig. 90, perfectly soft, otherwise it will make marks in the face of the hammer. This way of hardening and tempering will apply to all horseshoers' tools. When drawing out a pritchel, always strike the last two or three blows on the flat sides, but not necessarily to harden or temper.

How to Make a Horseshoer's Knife.

To make a horseshoer's knife, follow the instructions as given for making butcher knives mentioned in another chapter, with the exception that the horseshoer's knife must be bent to shape after the hammering is done while at a low red heat, and the temper drawn to a dark blue.

How to Dress a Vise.

It is a common occurrence when entering a blacksmith shop to find the blacksmith doing some work with the vise and at the same time blaming it for not gripping the work firm enough, on account of the teeth being worn smooth in the face and the work turns or slips when in the vise. But if the blacksmith knew how he could repair the vise in a couple of hours' time and make it grip as firm as when the vise was new. In order that the job may be performed successfully and with ease, follow these directions: Take the vise apart and place a jaw in the fire. If the jaw is worn very badly on the corners heat the jaw to a yellow heat and then forge to the proper shape. If the vise is not worn badly, just heat enough to draw the temper. After both jaws are dressed place the vise together to see if the jaws fit, as they should be perfectly level and straight. Now have a very sharp cold chisel and cut in the teeth at an angle across the face, beginning at one corner and going to the other. Then come back cutting at the opposite angle, as Fig. 91. Be careful not to cut the teeth in too deep or too far apart.

To harden, lay the jaw on the fire with the face or teeth side up; heat slowly to a low red heat. Then turn over and heat the whole face to a very even cherry red, but be careful not to overheat the corners and also bear in mind to have the center of the face as hot as is necessary to harden. Then plunge into the hardening bath and cool off entirely. Polish the face bright, and to draw the temper have the surface of the fire perfectly flat without any blaze; lay the jaw of the vise on the fire, teeth side up, and allow the temper to draw very slowly, without blowing the fire, with the exception of just enough to keep the fire from dying out. Allow

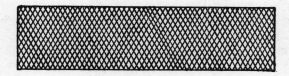

Fig. 91. Showing how teeth should be cut in a vise.

the temper to draw to a dark blue, then cool off, and the vise will hold the work very fast and firm, being equal to any new vise.

If the jaws of the vise are very square the teeth may be cut in cold without heating the vise at all, and which will answer the purpose very well in a temporary way. Make a cold chisel for cutting hard metal (after the directions as are already given in another chapter of this book) and cut in the teeth after the method as illustrated, being careful to keep the cutting edge of the chisel well ground.

Sharpening Plowshares.

There is a great amount of this work done throughout the country, especially in prairie States or Provinces and which in certain seasons of the year forms a great part of the country blacksmith's work, but there are very few blacksmiths who understand this work as it should be understood in order to give the farmer the best results.

The style or shape of the share varies according to the land which is to be plowed, as the land may be so-called stony, sandy or clay land.

The first and foremost point in being able to sharpen a share that will give good results is to have a thorough knowledge of the nature and the working of steel and being without this knowledge is the cause of shares breaking, bending or having so-called water cracks in them.

When sharpening a share have the hammer face very smooth, with the corners a little rounding, so as not to leave any deep marks in the steel, and also have the anvil face smooth, otherwise the share will give trouble by not cleaning when plowing in sticky or clay land, unless the marks are all ground out or the share polished after it is sharpened. When heating so as to draw the cutting edge, always heat the steel to a bright yellow heat, but always be careful not to overheat. When forging the cutting edge, hammer evenly and equally on both sides, but finish from the under side. By doing the last hammering from the under side it will have a very smooth surface on the top side of the share if the anvil is smooth; and always bear in mind to finish hammering when the steel is at a low black red heat, thus refining the steel and making it tough. Do not hammer the steel after it becomes black as hammering steel below a certain heat makes it flaky and brittle and easily broken. (This information will apply to farmers and others who try to sharpen their own shares by hammering them when cold.) Shares to be used in stony land should not be drawn out so thin as those which are used in sandy or land that is free from stones.

To harden a plowshare, heat the extreme cutting edge to a low cherry red (or just enough to harden) the full length of the share by moving back and forth through the fire, then plunge into clean cold water point first and in a vertical position. A plowshare sharpened and hardened after these directions will give unlimited satisfaction. But although the instructions given in hardening will prove most successful when it can be accomplished, there are times when a share cannot be hardened the whole length of the cutting edge in the blacksmith's fire. This will apply to very large and long shares, so when hardening very long shares do not try to harden the whole cutting edge, but harden only the point of the share from **a** to **b**, as illustrated in Fig. 92, a long share hardened

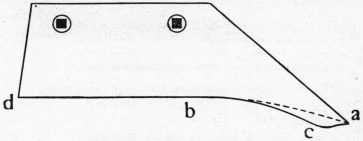

Fig. 92. An ordinary plowshare.

after these directions will give good results if the whole cutting edge has been forged and hammered after the instructions formerly mentioned.

When sharpening plowshares in order that they will give good results and cause the plow to run level, make the point a little rounding, as marked **c**, Fig. 92, otherwise if made square it would gather up long grass and cause the plow to run out of the ground. Also have the cutting edge in a straight line from the wing **d** to the point **a**, so that if placed on a level floor the cutting edge would come in contact and rest evenly on the floor, the entire length of the share. Plowshares for breaking plows are made with a very sharp point as illustrated by dotted lines in figure.

How to Make Square Holes in Plowshares.

A great many blacksmiths when making plowshares have more or less trouble to make proper shaped holes, in order that the bolthead will fit properly and not turn around when the share is being tightened up firmly to the plow head. To do this work correctly after the hole is drilled and countersunk, and the hole is to be made for a square cornered plow bolt, make a square drift or punch as Fig. 93, having it just

Fig. 93. Punch for squaring holes in plowshares.

a trifle larger than the square of the bolt head, also have the cutting face perfectly square and flat, it should be a little smaller back of the cutting face to give clearance. Harden the punch and draw the temper to a blue.

To make the hole square, place the share on the anvil holding it firm, then punch the hole as already drilled, from the opposite side of the countersink. By following this method and holding the punch directly and evenly over the hole, the corners will be cut out leaving the hole very square and neat.

CHAPTER XI.

How to make a harnessmaker's knife—Butchers' tools. Railroad tools.

How to Make a Harnessmaker's Knife.

To make a harnessmaker's knife will try the skill of the expert to its full extent, leaving aside an amateur steelworker, but if the amateur will follow the directions very closely he will meet with the best of success. To make one, use sheet cutlery steel of 75 points carbon and 3-32 of an inch thick, but the size of the steel to be used will depend on the size of the knife to be made. The size of the knife depends a great deal who is going to use it, as some knives are made as wide as 6 inches, while others do not exceed 4 inches, and in a great many cases smaller. After getting steel the proper width, draw an outline of the knife on the steel according to the size of the knife to be made, but bear in mind not to cut the full size of the knife as the cutting edge is yet to be drawn out, but cut out 3-16 smaller, as indicated by dotted line a, a, a, Fig. 94. If the steel is not long or wide enough to cut the full length of the shank, it may be cut as indicated by dotted line at b and then drawn out to shape, and in extreme scarcity of steel, the shank may be welded on at c, but this is a difficult job owing to the steel being so wide and thin, but it can be done with care.

After being cut to shape from the steel and the shank formed, forge the cutting edge of the knife by taking a deep cherry red heat, and commencing to hammer at the point marked f, hammering equally from both sides until all the cutting edge is drawn out,

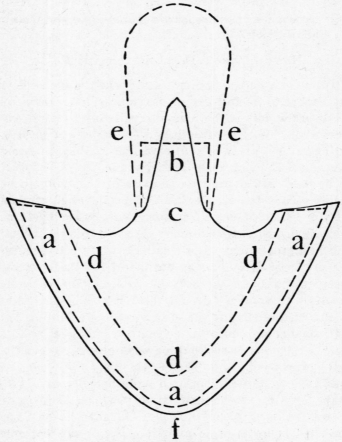

Fig. 94. The harnessmaker's knife.

although it may be necessary to take two or three heats before all the cutting edge is drawn out.

Now to refine and pack the steel, get the knife to a very dim red, then hammer evenly on both flat sides of the whole cutting edge; now heat again to a very low heat and then hammer, but this time do not hammer directly on the cutting edge, but instead go back from the extreme cutting edge ¾ of an inch, as indicated by d, d, d, which will take the strain out of the steel. Now anneal the knife by heating to an even blood red heat all over the knife (except the shank), then pack in lime or ashes, allowing it to remain there until perfectly cold, which will also help to take the strain out of the steel, afterwards filing the edges smooth. To harden, heat the knife to a low cherry red one inch back from the cutting edge in the blaze of the fire, at the same time moving the knife in a circular motion in order to heat the whole cutting edge very evenly (and be careful to avoid heating the cutting edge in streaks), then plunge the whole knife in the hardening bath from a vertical position, and cool off. If the cutting edge was forged and hammered very evenly and hardened evenly, the knife will come from the hardening bath perfectly straight, but otherwise there will be a crook in it, however polish one side of the knife as far back as it is hardened, and draw the temper very slowly over the fire, moving the knife in a circular motion back and forth to insure a very even temper, draw the temper to a dark blue (which will apply to all leather cutting tools), but before cooling off, should there be a crook in the knife, take it out by straightening it on the anvil with the hammer, then cool off and the knife is finished. Should it be impossible to take out all the crook with the hammer after the temper is drawn, a certain amount can be taken out and the knife made perfectly straight when grinding, but if the knife crooks a great deal when hardening, yet not enough to crack the steel, it will have to be heated again, straightened and annealed. However, if these directions are followed closely, there will be no danger of the knife crooking of any account, and the knife will hold a very keen edge on the hardest of leather. The dotted lines at e, e, e, represents an outline of the handle. To enable the shank to stay firm in the handle it should be feathered with a very sharp cold chisel, as illustrated.

How to Make a Butcher's Steel.

Butcher's steels are principally made by machinery, as the knowledge for making a good butcher's steel by hand is unknown by most of the blacksmiths or toolmakers, although a steel properly forged and hardened by hand will outwear any that can be purchased in a hardware store. To make one, take a piece of round tool steel say ⅝-inch diameter, after fullering in at a, Fig. 95, to form the shank and also fullering a

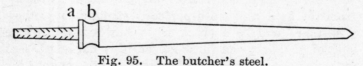

Fig. 95. The butcher's steel.

little at b, forge the steel tapering and very round. To put in the teeth, make the steel firm by putting the shank in a vise, now have a coarse sharp file, place it square across the steel, press hard and draw lengthwise of the steel from end to end and equally all around, two files will be found more convenient than one, on account that the teeth can be put in two sides of the steel at once, which is done by holding the ends of the files in the hands with the steel between the files.

Forge the shank long and square and feather it (as illustrated in Figure 95), so that it will stay firm in the handle. To harden, see instructions for hardening tools with fine projections.

Hardening Tools With Fine Projections.

To harden tools of this description, I will select a butcher's steel, as the teeth are very fine. Should this tool be heated to harden in the open fire as other tools, the teeth will lose their fine cutting edges and which is especially required on a steel in order to sharpen knives. To overcome this difficulty use the following compound: Take equal parts of wheat flour and salt, also a little water, then mix together to the consistency of soft mud. Have the steel perfectly dry and clean, then roll it in the compound covering the teeth of the steel well from end to end, then heat to a good cherry red the whole length of the tool (excepting the shank), then plunge vertically in the hardening bath, allowing it to remain there until perfectly cold, then clean off, but draw no temper, as the steel will not be too hard. Very fine files may also be hardened after this method and all other similar tools.

The Butcher's Cleaver.

A butcher's cleaver is made from flat steel ¼ inch in thickness, but the width of the steel will be according to the size of the cleaver to be made. When forging a cleaver, as Fig. 96, weld on an iron shank to the shape, as illustrated, so that it will be strong. The cutting edge of the cleaver is forged by drawing down the steel to a very short bevel equally from both sides while the steel is at a good yellow heat, then finish by hammering at a low heat on the beveled edges and backwards a short distance on the main body of the tool. To harden, heat the whole cutting edge backwards about ¾ of an inch to a cherry red by moving back and forth through the fire so as to heat evenly; then quench, afterwards drawing the temper to a light blue by moving it back and forth over the fire.

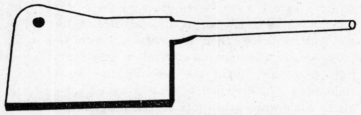

Fig. 96. The butcher's cleaver.

Butcher knives will be found fully explained and illustrated in another chapter of this book.

How to Dress a Railroad Pinch Bar.

Wherever railroad cars are to be moved by hand, it is often found necessary to take the pinch bar (as this is the tool principally used for this work) to the blacksmith to have it dressed or sharpened, but this work is very little understood by the average blacksmith and so the tool fails to give good results, it being too soft or otherwise the heel will break off.

To dress, heat the whole end of the pinch bar to a good yellow heat but not necessarily very far back, and then dress to the shape of Fig. 97. To harden, heat the face of the tool from heel to point to about 1 inch back to a very even cherry red, being careful not to get the heel overheated, then dip in the hardening bath to 1½ inches back at an angle as indicated by dotted line in figure. Now polish the heel and side bright and draw the temper, by placing the part between **a** and **b** directly over the fire, heat slowly and draw the point to a blue temper, but keep the heel as indicated by **c** cooled off by occasionally dipping it in water to the depth, as illustrated by dotted line (as the heel cannot be too hard when properly hardened) while drawing the temper at the point, then cool off

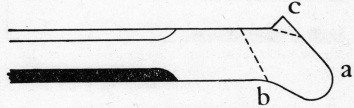

Fig. 97. Correct shape of punch bar.

and you have dressed a pinch bar that will give unlimited satisfaction.

There are different kinds or shapes of pinch bars, but the one as illustrated in Fig. 97 has been found by practical experience to be the best, as it is very strong. The one illustrated in Fig. 98 is perhaps more commonly made and used, but the fault of this pinch bar is, it is not heavy or strong enough at the point, consequently when moving heavy or loaded cars the point is very apt to bend or break. When hardening a pinch bar, as Fig. 98, harden the heel only, as the point will break if hardened or tempered. If at any time it is necessary to put a new heel on a pinch bar, upset the steel to the shape of Fig. 99, then forge to shape. The illustration as Fig. 99 will also apply when making new pinch bars.

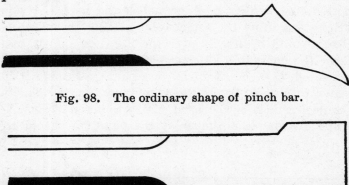

Fig. 98. The ordinary shape of pinch bar.

Fig. 99. Showing steel upset to forge heel on punch bar.

The Spike Maul.

When building or repairing railroads, the spike maul or spike hammer is greatly used, its principal use being to drive spikes in the ties. To make one, take a piece of 2-inch square steel and after punching the eye forge to the shape as illustrated in Fig. 100, as will be

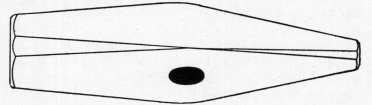

Fig. 100. The spike maul.

seen one end is drawn down very small to about ¾ of an inch across the face, both ends are dressed as an ordinary hammer. To harden, heat the large end of the spike maul first to a cherry red about ¾ of an inch back from the face, having the center of the face at an even heat with the outside or corners in order that it will harden properly, then dip in the hardening bath to about 1½ inches deep. Then polish the face and allow the temper to run down to a light blue. If there is not enough heat in the spike maul to drive down the temper it can be drawn by holding the end over the fire, and slowly and continually turning it around until the temper is drawn to the desired color. Harden and temper both ends the same way, but be careful not to draw the temper in the end which is already hardened and tempered while heating to harden the second end. However, if the large end is hardened and tempered first, there will be no danger of the temper drawing in the other end, as the small end can be heated so much quicker and so the heat has not time to run to the large end enough to draw the temper, but in cases where there is danger of the temper drawing, cool off the tempered end in water. A spike maul is not so apt to get hollowing in the center of the face as an ordinary hammer, but, instead, is more apt to break off at the corners, so when dressing make the corners rounding, but not too much.

The Claw Bar.

In railroad construction, the claw bar is very extensively used, its principal use being to pull spikes. Dressing claw bars when badly broken is somewhat of a difficult task and requires skill to forge them to the proper shape. To dress a claw bar, as shown in Fig. 101, when badly worn or broken, heat to a deep yellow heat, then forge to shape, as illustrated by side view in figure, then close the claws, as a, a, front view in figure, to within ⅜ of an inch apart at the extreme ends. Now have a fuller as Fig. 103, which should be

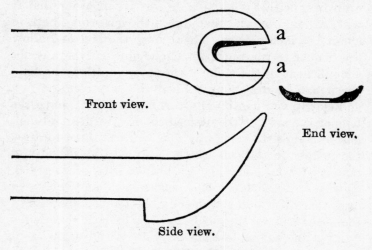

Fig. 101. Illustrating claw bar.

⅛ wider than the body of a railroad spike, and drive the fuller down between the claws. This will straighten the claws and bring them the right width apart. Now

Fig. 102. Side view of opposite end of claw bar.

have a small gouge and gouge out the claws from the front side, the extreme ends should be very thin (as shown in end view of figure, in order to go under the head of a spike when pulling it), but should gradually become thicker back from the extreme ends.

Claw bars must not be hardened or tempered, or they will break very easily when pulling a hard spike. The opposite end of a claw bar is generally made with a

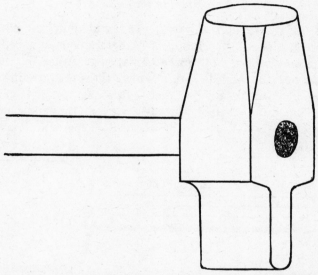

Fig. 103. Illustrating kind of fuller used, when dressing claw bars.

bent chisel point as Fig. 102, which is used sometimes to loosen the spikes before pulling. There are also many different shapes of claw bars, but the one illustrated is the principal one used, and they are all dressed after the same method or principle.

CHAPTER XII.

Miscellaneous tools—Case hardening.

The Bricklayer's Set.

A bricklayer's set has a very wide cutting edge, while the shank or handle is ¾ octagon or square, to make one, take a wide piece of flat steel fuller in, then draw out the handle, after which the cutting edge is forged. But in case a piece of steel, as just mentioned, cannot be had, the only way to forge the cutting edge (which should be about 3 inches wide) is by

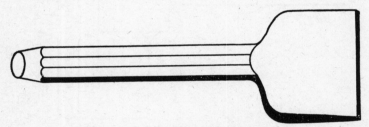

Fig. 104. The bricklayer's set.

upsetting a piece of octagon or square steel as the case may be, then flatten out until wide enough. To put on the cutting edge file only from one side, the other side being left perfectly square and flat.

Harden as any similar flat tool and draw the temper to a dark blue. All tools for cutting brick should be tempered to a light blue.

How to Harden and Temper Wire Nippers or Pliers.

Heat the jaws back a little past the cutting blades, as indicated by dotted line **a**, Fig. 105, to a very even cherry red, then dip into the hardening bath to dotted line **b** above the rivet. Now polish the upper side

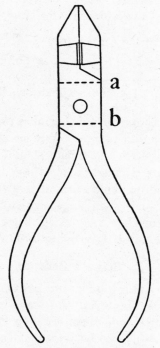

Fig. 105. The wire nippers.

bright and draw the temper over the fire very slowly and evenly to a light blue, making sure that the cutting edges or teeth are properly tempered. These instructions will apply to all similar tools.

How to Make a Razor.

To make an ordinary razor use steel 3-16 by 7-16 of about 75 points carbon. After the shank is fullered in a little for the finger hold, then forged to shape, the blade is formed into shape by bending the steel a little edgewise, afterwards being forged, hammered and hardened, as is explained in making butcher knives

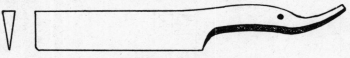

Fig. 106. The razor.

(which will be found in another chapter of this book), draw the temper to a purple. A razor is hollow ground after being tempered and which should be done by an expert, if a razor is made after these instructions and the hollow grinding done without drawing the temper, it will hold a very keen edge, which will equal any razor manufactured. End view, Fig. 106, illustrates shape of razor blade before being hollow ground.

To Make a Scraper.

A scraper for taking off paint, grease, etc., off boiler plate or any other material and leave a bright smooth surface, is chiefly made from octagon steel, the size of steel used according to the width of scraper required, although for an ordinary scraper ⅝ octagon will do. To make one, forge the steel perfectly flat ⅛ thick and about 3 inches in length and 1 inch wide. The end of the tool is left perfectly square, the scraping or cutting edges being the corners, which are ground very sharp.

To harden, heat to about 1 inch back from the scraping edges, then quench in the hardening bath and cool off the whole tool entirely. Draw no temper as the tool is required to be very hard; it will give excellent results if properly forged, hammered and hardened.

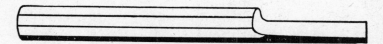

Fig. 107. Showing scraper for boiler plate, cast iron, etc.

Hardening Jaw of Pipe Vise.

To harden a jaw of a pipe vise, heat all the teeth to a very even cherry red or just enough to harden, then quench the whole tool edgewise from a vertical position in the hardening bath and cool off entirely. Polish one side bright and draw the temper to a dark blue by plac-

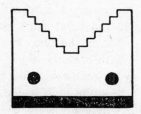

Fig. 108. Jaw for pipe vise.

ing the jaw on a heated iron plate which should be a little wider than the jaw, in order that the jaw may be tempered evenly. These instructions will apply to all tools for holding pipes, clamps for holding bolts and all similar tools.

Hardening and Tempering Blacksmiths' Bolt Clippers.

A good set of bolt clippers is a tool prized very much by the general blacksmith, and yet very few blacksmiths are capable of repairing them properly when they get out of order, the greatest trouble lying in the hardening and tempering.

After the clippers are dressed and the cutting edges made to fit properly and closely together, heat the whole cutting edge to a very even cherry red, then quench in the hardening bath from an upright position to about one inch from the cutting edge. Polish the cutting edge bright and draw the temper slowly and evenly over the fire to a light blue. These directions for hardening and tempering bolt clippers apply to the kind that are used principally nowadays, which have a long shaped blade and by dipping into the hardening bath (after the fashion of hardening a cold chisel) one inch back from the cutting edge, will enable the temper to be drawn more accurate and evenly with no danger of the temper running out at any part of the cutting edge if the least care is exercised when drawing the temper over the fire. However, some bolt clippers are made with a short blade which is held in place by a set screw or some other contrivance, the knife or blade not exceeding one inch in length. In a case of this kind heat the whole blade to a very even cherry red, then quench the whole tool in the hardening bath and cool off entirely, afterwards drawing the temper on a piece of hot iron or by holding it over the fire.

Bolt clippers are made exclusively for cutting iron bolts or rivets and must not be used to cut cast steel, if used on cast steel the clippers will lose their sharp cutting edges or will break.

Tools for Punching or Gumming Cross Cut Saws.

A punch and die for gumming cross cut saws are made a great deal after the same principle as a punch and die for punching teeth in stone cutters' chisels, with the exception that the saw tools are not beveled off, but instead are left perfectly flat, the hardening and tempering being the same. See "punching teeth in stone cutters' tools" in another chapter of this book. All punches for saw sets, after hardening, should be tempered to a light blue.

The Scratchawl.

A scratchawl for scratching or marking cast iron, boiler plate, etc., is as a rule made from small round steel, the point being drawn out very long and thin. To harden, heat to about ½ inch back from the point, but owing to the fineness of the tool be very careful not to overheat the extreme point, then quench and cool off entirely, draw no temper, as the point is required to be very hard. Most mechanics who have use for a scratchawl prefer the opposite end flattened and bent to shape as Fig. 109.

Fig. 109. The scratchawl.

Hardening and Tempering Circular Blades of Pipe Cutter.

To harden circular blades of pipe cutter, heat the whole blade to a very even cherry red heat, then quench the whole tool and cool off entirely. Afterwards draw the temper to a dark blue on a piece of heated flat iron. Should there be a great many of these tools to be hardened at once and there is no heating furnace in the shop, place a piece of flat iron on the surface of the fire, heat it to a deep red heat, then place the blades on it, as the blades are of a flat shape it will not take long for them to heat hot enough to harden. Place 5 or 6 of the blades on the heated plate at one time, but watch carefully and keep turning them over for fear some of them should become a little overheated or heated in streaks. After quenching draw the temper also on a hot iron.

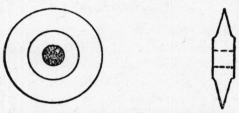

Fig. 110. Flat and end views of circular blade for pipe cutter

Heating a Tool According to Its Shape.

When heating, to harden tools of an irregular shape as an eccentric ring, Fig. 111, the heavy or thick side

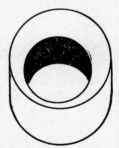

Fig. 111. An eccentric ring.

should be heated first, then allow the thin part to come up to the heat gradually so as to avoid unequal contraction when hardening. When quenching plunge the heaviest part of the tool into the water first.

Making, Hardening and Tempering an Alligator Pipe Wrench.

To make an alligator pipe wrench, take a piece of flat steel the size according to the size of wrench to be made, heat and then fuller in as a, a, Fig. 112, afterwards drawing out the handle b, now cut off the four corners, as illustrated c, to shape as indicated by dotted lines with a thin splitting chisel which will give the shape of the wrench. Now punch a small hole in the wrench at d and cut out the part e as dotted lines. If the wrench is made where there is a machine shop the teeth can be put in with a planer, but if made in an ordinary blacksmith shop the teeth will have to be filed in. The teeth can be put in one or both jaws, as may be desired. Fig. 113 shows the completed wrench with teeth in one jaw.

To harden, heat the jaw (having the teeth) enough to harden to dotted line a, Fig. 113, then quench into the hardening bath to dotted line b, polish one side bright and draw the temper over the fire to a dark blue. Should both jaws of the wrench have teeth it can be hardened and tempered after the same method, but if the wrench has teeth in only one jaw, it is not necessary to harden or temper the jaw having no teeth. These directions will apply to all kinds of alligator wrenches or similar tools.

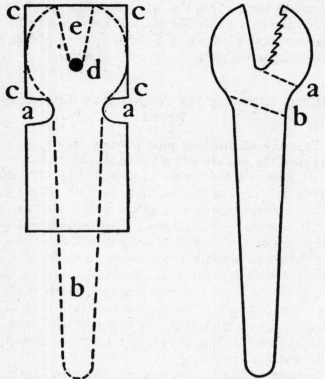

Fig. 112. Showing how alligator pipe wrench is made.

Fig. 113. The completed wrench.

Hardening and Tempering Pruning Shear Blades.

If the blade is short, heat the whole blade to an even cherry red heat, then quench in the hardening bath about an inch back from the cutting edge and in an upright position, afterwards polish and draw the temper over the fire to a light blue. Should the blade be long, say 6 inches or more, harden and temper as a butcher knife, mentioned elsewhere in this book.

The Center Punch.

A center punch for marking or centering holes that have to be drilled in iron, steel, etc., are drawn down to a very sharp point as shown in Fig. 114. After hardening, allow the temper to draw to a dark blue,

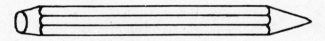

Fig. 114. The center punch.

which will do for punching all ordinary material, but for punching very hard metal the temper must be regulated accordingly. See instructions as is given in tempering a cold chisel in another chapter of this book.

The Nail Set.

A nail set for driving nails deep into the wood is generally made from ⅜ octagon or square steel. The end

Fig. 115. The nail set.

for striking upon the nail is tapered to ⅛ at the point. Harden not less than ¾ of an inch back from the point and draw the temper to a light blue.

Hardening and Tempering Steel Stamps.

Stamps for lettering or marking cold iron, steel, etc., are hardened as any ordinary tool by being heated and quenched about one inch from the stamping end, and afterwards drawing the temper to a purple. Stamps for marking hot iron or steel will be best tempered to a light blue. When stamping cold material, be sure to always have the stamp perfectly level and firm on the material to be stamped, otherwise the tool will be apt to break.

Making a Gouge.

To make a gouge for cutting hot iron or steel, it must be first made as an ordinary hot or splitting chisel, but the cutting edge should be left a little wider than the body of the chisel, then it is placed over a bottom swage and while at a cherry red heat take a fuller and place it exactly in the center of the chisel and directly over the center of the swage, then strike the fuller a good blow or two with the sledge, which will set or force the chisel down into the swage and form the gouge. Bear in mind that a certain size of swage and fuller must be used according to the size of gouge to be made, for example, and to have the best success, supposing a gouge is to be made to cut a circle of one inch, the swage must be one inch and the fuller $5/8$ of an inch in size. This method will also apply to making a carpenter's gouge. A gouge for cutting cold iron or steel must be left thicker than one made to cut hot material and which will require a smaller size of fuller when making one. Also bear in mind to have the steel at an even cherry red heat (but no hotter) when bending a gouge to shape, otherwise should it be bent while at a white or high yellow heat, the hammering which is done at a low heat (before bending) is all taken out and it will never hold a keen cutting edge or otherwise if the gouge should be bent at a very low or black red heat there will be strains put in the steel which will cause the gouge to crack while hardening.

To harden a gouge, follow the directions given in hardening and tempering a blacksmith's hot or carpenter's chisel, which will be found fully illustrated in other chapters of this book.

Hardening and Tempering Carpenters' Augers That Have Come Through a Fire.

Although augers are made, hardened and tempered entirely in tool factories, there is often a case when an auger has simply lost its temper and become soft in a fire by the burning of hardware stores, etc., although otherwise the auger is not damaged in the least and which can be made as good as new (unless already overheated) by the following method: Heat the auger very carefully in the blaze of the fire, making sure that the cutting edges and point are heated very evenly to a cherry red, then quench into the hardening bath about one inch back from the cutting edges, polish bright and draw the temper very carefully over the fire (having no blaze) to a dark blue. These instructions will apply to all augers, brace bits and all similar tools for boring wood.

Case Hardening.

Case hardening is a process that iron or soft machinery steel is put through so that the outside will be made very hard, while the centre still remains in its soft state. Case hardening is a very useful treatment, as certain parts of machinery are to be very hard in order to stand the wear, and iron or soft machinery steel can be made to give very satisfactory results when case hardened properly. For a great number of purposes machinery steel tools will take the place of tools made from cast steel, and is less expensive as the stock is cheaper and the tools are much more easily made and will last just as long when properly treated.

When case hardening parts or ends of tools, such as set screws, the process is this: Heat the end of the set screw to be hardened to a bright cherry red, then roll the heated end into powdered prussiate of potash and return to the fire and heat to a bright cherry red, then plunge into cold water and cool off entirely, when it will come out of the water presenting a white appearance and will be found to be very hard if tested with a file. This method will apply to all small tools that are to be case hardened all over, but not to a great depth.

When case hardening a great many tools at once where the whole surface is to be case hardened and to a great depth, pack in an iron box with any case hardening compound, such as granulated or charred bone, charred leather, charcoal and potash. But in packing the pieces or tools in the box be sure that they do not come in contact with the surface of the box or with each other, but keep at least ½ inch apart by packing the case hardening compound between. The box should be made air tight, and then placed in a furnace or heating oven which must be left there long enough according to the depth that the contents are to be case hardened. If the furnace is kept at a bright cherry red heat, the contents of the box will be case hardened to the depth and rate of 1-16 of an inch per hour. Afterwards the contents are taken from the box and quenched immediately in cold water and cooled off entirely. When quenching pieces or tools to be case hardened, the process is the same as cast steel, for example, take a flat piece of iron or soft steel 6 inches square by 1 inch thick it must be quenched by plunging into the water edgewise and from an upright position, also long slender pieces (for explanation 1 inch square and 6 inches long) must be quenched from a perfectly upright position and never at an angle, otherwise if the pieces are quenched at an angle they will be apt to warp.

CHAPTER XIII.

General information, pointers and ideas in reference to steel work and toolmaking—Conclusion.

The Correct Meaning of a Cherry Red Heat.

We often hear blacksmiths and other mechanics when in conversation about steel saying such a tool or piece of steel should be heated to a cherry red to harden or temper, but although their advice may be quite true how many are there who know the correct heat and meaning of a cherry red heat? If every blacksmith and other mechanics who claim to know all about steel were judged according to the class of tools they make, then I am afraid there would be a great many who do not understand the correct heat or meaning of a cherry red heat. I personally know some blacksmiths and tooldressers who will tell me they know a great deal about steel and its nature, who, when they are hardening a piece of steel will often heat the steel to a bright yellow heat when they think it is a cherry red, while others again will not have the steel heated above a dull red heat, consequently the steel is heated too much or not enough to harden. But for the benefit of those who wish to know and are anxious to learn the real meaning of a cherry red heat, I will explain. A cherry red heat is the lowest heat at which a piece of steel containing 75 points carbon will harden successfully. But when hardening a piece of steel containing 100 points or 1 per cent of carbon a lower heat than a cherry red will do to harden it, and it should always be well remembered that the lowest possible heat that any steel will be sure to harden at, so much better will be the results when the finished tool is put into use, consequently the steel (to have the best results) must be heated to harden according to the carbon it contains.

Heating to Harden According to the Size of the Tool.

When heating to harden large or heavy tools, it should be remembered the heat should be a shade higher than that used to harden small or thin tools as the water will act much quicker on a thin piece of steel than a thick piece. Thick heavy tools will cause a great amount of steam, which has a great tendency to hold the water back from the steel, and to more fully explain, if two pieces of steel are taken to be hardened, the larger piece being 3 inches square, the smaller being ½ of an inch square, but both pieces containing the same amount of carbon. Now if the smaller piece is heated to the lowest possible heat, that it will harden at successfully clear through the piece. Then heat the larger piece to exactly the same degree of heat, and it will be found after hardening upon close examination that only the corners are hardened.

When quenching the ends of large tools or wholly, keep the water well agitated and so help to keep the water cooler next to the steel. If it is possible to have an overflow pipe attached to the hardening bath and another one to flow in at the same time, it will give good results when hardening large pieces as the hot water will continually flow away from the steel.

Charcoal for Heating Steel.

A saying that I have heard a great many times is, always use charcoal to heat steel, while another saying is, steel is tougher when heated in charcoal. It has been found out by practical and scientific experience that sulphur is one of the greatest enemies to be avoided when heating steel, and while charcoal is free from sulphur it is the only advantage connected with charcoal to heat steel by. But as to toughen steel by heating it in charcoal, is a saying entirely without foundation, as there is nothing that will toughen steel except by hammering it at a certain heat and hardening it at the proper heat.

In reference to heating steel by the use of charcoal, it does not matter what kind of fuel is used so long as it is free from sulphur and the necessary heat can be obtained, whether it be charcoal, coke, coal, bark or corncobs. Now, reader, never adopt any old saying or process unless you have found out by experiment or trial that it is true. Some mechanics are too ready to pick up quack theories by having heard some one say so, and consequently when the advice is put into practice the result is failure. Reader, there are too many mechanics who belong to this class, you must belong to the class who do their own thinking, which is the sure road to success. Again, never turn a deaf ear to any one who has a suggestion to make, no matter who the man is or what kind of clothes he wears, the author has picked up some very valuable information in this way from some of the most illiterate men who only excelled in one point and which has been found out by experiment to be reliable.

The Scaling of Steel After Hardening.

The scaling of steel after hardening, as a rule, is never observed by the average steelworker, but to the expert it means a great deal. By the way the steel scales off a good steelworker is enabled to tell good or poor steel, also hard or soft steel, hard steel containing 100 points carbon if heated to a cherry red will scale off clean, leaving a white surface, while a piece of medium carbon steel of 75 points of good quality and heated to a cherry red will scale off in spots leaving a speckled black and white surface, but the scales left on will be very thin and light. But if a piece of steel of medium carbon and of poor quality be hardened, the scales left on will be of a thick and heavy nature, and steel very low in carbon and of poor quality will not scale off at all unless heated to a bright cherry red (almost yellow). This class of steel is worthless for making good tools. The temper of a tool may also be regulated by the way the steel scales off. For illustration, if a cold chisel (for ordinary work) after hardening should be partly scaled off, the temper should be drawn to an ordinary light blue, but should it scale off perfectly clean and white allow the temper to draw to a very light blue, bordering on a grey.

Sometimes when heating steel in a coke or smoky fire the steel will scale off as already mentioned, but instead of leaving a white surface it will present a very dark surface and unless observed closely it will be hard to tell whether the steel has scaled off or not. The scaling of steel, however, is only in reference to tools that are hardened after being finished at the anvil, as tools that are ground bright on a grindstone or otherwise machined will not scale off.

By the scaling of steel a good steelworker can tell if the tool has been overheated when hardening, as the surface of overheated steel will show a very bright white color, the best way, however, to learn the difference as regards the scaling off of a piece of overheated steel and of steel that was hardened at the proper heat, take two pieces of steel from the same bar of good quality and medium carbon, then heat one to a cherry red and the other to a deep yellow or white.

Quality and Quantity.

If tools can be made or repaired very quickly and in great quantities by a toolsmith or any mechanic, producing at the same time excellent quality, it will be a great saving of time, but there are very few who have such good ability. But to the ordinary toolsmith or steelworker I wish to give this advice, "Let quality at all times be preferred to quantity," and always see how well you can make a tool before seeing how fast you can make it, speed will naturally come but quality must be practiced in the beginning. For illustration, we will take two different toolsmiths who are making cold chisels, one may be able to make in his own way 50 chisels while the other man is only making 25, but the one who makes 25 is having a much easier time and is always up to his work, although he has as many mechanics to keep at work as the man who makes 50, but how is that the reader asks; it is because he prefers quality first. The other man is working as hard as he can, he never catches up to his work because he prefers quantity first. The slow man in action but not in workmanship makes one chisel that does as much work as three chisels made by the swift man, consequently every chisel made by one toolsmith does its work while almost every chisel made by the other toolsmith is continually breaking, bending or being too soft.

Quick Methods of Working.

When making or dressing a great many tools of the same shape and used for the same purpose, first of all consider the quality, then consider quantity and the quickest way of doing the work. For example, supposing 25 cold chisels are brought at one time to the toolsmith to be dressed, do not dress, harden and temper them one by one, but instead dress or draw out all the chisels (using two pair of tongs and so keeping one chisel heating in the fire while the other one is being drawn out) before hardening and tempering them.

When hardening the chisels have the surface of the fire flat, then lay about 4 of the chisels on top of the fire and heat very slowly, as they become hot enough to harden quench deep into the hardening bath. After polishing lay the chisel down on the forge or in some very convenient place near the fire (so that the toolsmith may be able to watch the chisels that are on the fire and also watch the temper drawing on others) and allow the temper to draw on its own accord, if there is plenty of heat left in the chisel back from the hardened part although dipped $1\frac{3}{4}$ inches into the hardening bath, the temper will often draw showing a light blue color $1\frac{1}{4}$ inches back from the cutting edge, but if the temper does not draw quite to the necessary color on its own accord hold it over the fire. When heating the chisels for hardening after this method let the helper (if you have one) keep the chisels placed on the fire in order that they may be continually heating, as others are taken away to be hardened, "but do not trust the helper" to heat them to the right heat, as the toolsmith must watch the hardening heat himself as he is responsible for the quality of the tool. With a good helper, the author has dressed, hardened and tempered from 15 to 20 flat cold chisels per hour after the method already explained, and every chisel guaranteed to give the best of satisfaction by chipping as hard steel, as the chisel is made from.

When dressing granite cutters' tools, keep 3 or 4 in the fire gradually heating and as one becomes hot enough to dress, dress it and return to the fire to heat for hardening (as granite tools do not require to be heated very far back from the cutting edge, when hardening they can be heated very quickly), when hot enough to harden, quench it deep into the hardening bath, then rub it across the sand board to brighten, afterwards placing it on a piece of sheet iron or tin attached to a tub of water (not the hardening bath) and allow the temper to draw on the tool by its own accord, as granite tools do not require much temper, very little heat left in the tool after quenching will be sufficient to draw the temper to the desired color, and as the temper becomes drawn in a tool push it off into the water to cool off. After a little practice and having all the tools, etc., placed in a very convenient position, 50 to 60 tools such as points, chisels, and small drills can be dressed, hardened and tempered, in one hour.

When hardening and tempering a great many small machinist's or riveting hammers, harden both ends at once by holding the hammer in an upright position and reversing the ends of the hammer back and forth into the fire (thus heating the ends hot enough to harden, but keeping the eye of the hammer soft), then plunge the whole hammer into the hardening bath and cool off entirely, then polish bright. To draw the temper, heat a large iron block (say 6 inches square and 4 inches thick) to a deep yellow heat, now place the hardened hammers across the corners of the heated block, having the eye directly on the block and so allowing the ends to project out from the heated block as shown by dotted lines, Fig. 116. By this method of

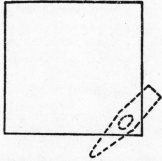

Fig. 116. Illustrating quick way of tempering small hammers.

tempering small hammers, the temper will draw in the eye first if the hammer is turned over occasionally, afterwards the temper will draw in the ends, by keeping 4 hammers on the block at once they can be tempered very quickly. If one end of a hammer should draw the temper before the other cool it off, but not any further back than the eye, then place on the heated block again until the temper is drawn in the other end.

When forging tools, keep your hand tools well arranged in a very convenient place close to the anvil and so be able to put your hand on the tool at once (which is required) and save lifting or moving a great many other unnecessary tools.

Cracks in Steel.

Anyone who has had any experience working steel or making tools has noticed cracks in the steel as shown in the cold chisel, Fig. 117. These cracks are often called "water cracks" and some will say the water was too cold, but the real causes are having the steel too hot when hardening and unevenly forging the tool at too low a heat, and if the tool has been forged at a very low heat then overheated when hardening, cracks in the steel are almost sure to be the result. Any tool that has a crack in it, no matter how small, will break very easy when put in use.

Slighting Tools.

Any toolsmith who wants to do good work in order that he may gain a good reputation cannot afford to be careless or slight the tools he is to make or repair, but should at all times do the work to the best of his ability. When dressing cold chisels as the one illustrated in Fig. 117 (or any similar tool) do not attempt

Fig. 117. Showing cracks in cutting edge of a cold chisel.

to dress it by leaving the cracks in it, but cut off the end of the chisel as far back as the cracks are visible in the steel.

A steelworker should never get in the habit of doing much talking while a piece of steel is heating in the fire, and more especially if the steel should be heating to be hardened, a great many expensive tools have been entirely destroyed by doing too much talking, thus causing the toolsmith to forget his work. If you have a lot of talking to do which is apt to draw your attention away from the fire, take the steel out of the fire and lay it to one side until you are through with the conversation, as there is no work which demands closer attention and greater care than successful toolmaking. Also, do not get excited when working steel, as no one can work steel successfully if he is of an excited and nervous nature and never attempt to make tools when under the influence of strong drink, to work steel successfully the steelworker must have a clear brain and be patient, careful and have a quick and decisive judgment.

The Result of Being a Successful Steelworker.

The keynote of being a successful steelworker is "economy," and any large manufacturing company having in its employ an expert steelworker will be a great amount of money ahead at the end of a year, this result caused by the saving of steel and also labor. For example, I have been in large manufacturing establishments where they had a poor toolsmith in their employ, and the results were, a great many mechanics were losing time going back and forth to the tool fire in order that they might get a tool to do good work. I have known machinists and others to go to the tool fire 3 or 4 times (while once is enough) before they would get a tool to do its work, and in some cases they never got a first-class tool.

Not only is the result of being a successful steelworker or toolsmith a great saving to the firm or company he is employed with, but a saving of a great amount of unnecessary work and trouble to himself. A first-class toolsmith will keep a much larger gang of mechanics at work, when placed beside a toolsmith whose knowledge concerning steelwork or toolmaking is very limited.

Hardening Tools That Are Forged By Another Mechanic.

No blacksmith or steelworker should assume the responsibility of hardening tools (especially if expensive) that are forged by another mechanic, for if the tool should crack when hardening (or in any other way not prove satisfactory) the blame will fall on the mechanic who hardens the tool, although the one who deserves the blame is the man who forged the tool, as he left strains in the steel by irregular heating and irregular hammering. It does not signify how expert the hardener may be, as the steel is still bound to crack if not forged properly.

Sayings and Ideas of Mechanics in Reference to Steel.

Steel is one of the greatest and favorite subjects for discussion among mechanics, and the consequence is a great many different theories, sayings and ideas arise in their minds as to what they think is the best way of working it, and the toolsmith is in a position to hear the idea of each mechanic when wanting a tool made or repaired. But the toolsmith must do as he thinks right as regards the working of the steel, but as to the shape of the tool he must follow the mechanic's instructions. For example, I will give a few of the sayings and ideas which I have gathered up at the tool fire, although other sayings are mentioned in this book. Some mechanics have a habit of trying a tool with a file before using it, and not long ago I made a chisel for a machinist. After I had it made he took it away and in a few minutes returned with the chisel filed to an edge. He came up to me and said, "This chisel is no good because I can file it!" I asked him if he had tried it. He said, "No, there is no use of me trying it when I can file it so easy!" I told him to go and try it. He went and tried it and was convinced that the chisel would hold a good edge although he was able to file it. Now I wish to say to anyone who uses a cold chisel, although a chisel will chip cast steel and at the same time hold a good edge, it can be very easily filed to an edge with the use of a new or sharp file.

An idea with a great many so-called steelworkers is, they will say, it does not matter Low hot steel is heated (so long as it is not burnt) when hardening, if it is put in the water at a cherry red. Now I wish to say in reference to this idea, although the result will not be as bad, neither will the steel be so liable to crack as when quenching the steel at a deep yellow or white heat, but the steel will never be in a fine crystalized state or hold a keen cutting edge as a piece of steel heated to the proper hardening heat, which is mentioned so often in this book. Again, steel should never be held in the air to cool off after it leaves the fire when it is to be hardened, so bear in mind when hardening steel to quench it in the hardening bath directly after it leaves the fire, and have the hardening bath as close to the fire as possible and in a convenient place.

Forge steel at a low heat is an old saying of a great many, and likewise a great many blacksmiths believe it and so forge the steel at a cherry red heat. I have already mentioned about the dangerous practice and hard work of forging steel at a low heat, but for the benefit of any reader who is connected with the working of steel, I will give this information, trusting it will be well remembered. Forge cast steel in the beginning at a deep yellow heat, and if the steel is heated to a white heat (so long as it is not burnt) it will be all right, but lessen the heat as the tool nears the finished stage, all tools should be finished at a very low heat, especially all edged tools having a flat surface.

Never upset cast steel is another old saying of some steelworkers, and when they are asked the reason why they will generally reply, "It opens the grain and spoils the steel," and I have known some blacksmiths who refused to make certain tools, saying to the customer the steel was not large enough without upsetting and that would spoil the steel. Reader, upsetting spoils the steel is an old saying, but that is all there is to it, as there is practically no foundation in the theory and there is positively no grain in steel. Sometimes there happens to be seams lengthwise in the bar but that is a fault of the manufacturer. A piece of good steel may be said to resemble a piece of putty, which can be worked in any way and still produce the same results so far as the grain is concerned. The author has upset steel to three times its original size and when finished gave as good satisfaction as if it had not been upset at all. But to more fully explain, take a piece of octagon steel and make it into a chisel and when made the chisel is found to cut first class, now what is to hinder us from making a chisel on the opposite end. There is simply nothing at all, "and yet we are upsetting it, as drawing the steel out on one end is working the steel in the same direction as upsetting it in the other end. Now, reader, remember this: A cutting edge can be put on the side of the octagon bar and still stand as well as if put on at the end; also the steel can be upset, crooked or bent any shape or form and still hold a first-class cutting edge when properly worked.

Why Some Tools Are Soft When Put Into Use.

There are different reasons for tools being soft when put into use. However, the main reason is, the tool has not been hardened successfully as the steel was not hot enough when quenched into the hardening bath, and if the steel has not been hardened it matters not whether any temper is drawn or not the tool will be soft just the same. Another reason why a tool is often said to be soft by some mechanics when bringing it to the toolsmith to be repaired, is because the tool has been used on hard cast iron while the tool was not tempered to cut anything harder than ordinary cast steel, consequently the cutting edge turns over when coming into contact with the hard metal.

Still another reason, although the tool may be hardened and tempered properly, it is quite a common occurrence that the temper is drawn when grinding on an emery wheel.

Reasons Why Tools Break When in Use.

A few of the main reasons why tools break when in use are—overheating of the steel when hardening, and improper forging which will cause cracks in the steel, also by leaving the temper too high, making the tools too thin and using a poor quality of steel. An other reason for tools breaking is the result of putting tools to a use for which they were not intended. For explanation, I have had mechanics bring me broken cold chisels (and having made the chisels myself, I could certify that they were forged, hardened and tempered correctly) and when I would make inquiries to ascertain how the chisels were broken, the mechanic would say he was using them for wedges. Very often the toolsmith worries and so keeps himself in hot water when broken tools come to him to be dressed, as he thinks he is to blame because he thinks he did not make the tools properly. This is a great mistake on the toolsmith's part, for in a great many cases the mechanic who is using the tools is to blame, for if a tool is put into a use for which it is not intended or used carelessly it does not signify how well the tools are forged, hardened or tempered, they will break just the same, and for the benefit of every blacksmith or toolsmith who chances to get a copy of this book, I say follow the instructions closely concerning each tool, and then if the tools are broken when in use 9 cases out of 10 the fault will lie with the mechanic who was using the tool. When a tool is broken look at the fracture of the break, if it presents a close grain resembling a piece of glass the tool has been hardened properly. But, instead, should the break present a very coarse fracture resembling somewhat a piece of honeycomb, the tool has been improperly hardened by overheating, and in a case of this kind the toolsmith is to blame for the tool breaking. Tools used in very cold and frosty weather will break much easier than in warm weather, especially if used outside in the open air.

Necessary Tools.

I have often strolled into a country or general blacksmith shop and found the blacksmith trying to forge a piece of neat work, by using simply a hammer and the anvil. It is quite possible that a great amount of work can be accomplished with only a hammer and the anvil, but the work is limited and very often after forging the article as near the shape as is possible, the blacksmith will often wear out a new file by filing the article to the finished shape, while the money that he pays out for files would soon amount up enough to buy him a good outfit of anvil tools or pay him for the time it would take to make them. By having a good outfit of tools, a great many jobs can be done in half the time and give a much neater appearance. The most necessary tools required in the ordinary blacksmith shop, aside from a good anvil and hammer are tongs for holding different shapes and sizes, fullers and swages ranging in size from $\frac{1}{4}$ inch to 2 inches, a flatter, set hammer, a hot and cold chisel, and a hardy. Of course there are a great many useful tools that I could mention, such as are used in large and up-to-date shops, but in a small shop it would not pay to keep them all on hand. However, the ones already mentioned should always be kept on hand. Any blacksmith who is employed in a large machine, locomotive, or any large shop, should always be on the lookout to have as good an outfit of tools as anyone else in the shop, and so save borrowing from another fire. Although in some cases it is necessary to borrow, but when borrowing tools from another fire bear in mind to take them back as soon as possible, otherwise it may cause trouble.

Welding Compounds.

There are a great many different kinds of welding compounds, but the kind that is most extensively used and most commonly known is borax. The borax should be crushed to a fine powder to have the best results, and if wrought iron drillings (that are very fine and free from oil) be mixed it will increase the welding qualities of the borax by causing the steel to unite more readily.

Although all the instructions that I have given in the previous chapters (in reference to welding) is with the use of borax, there are other compounds which I have used with great success. A welding compound that I will recommend to the reader is the Climax, manufactured by the Cortland Welding Compound Co., Cortland, N. Y. This compound is very valuable when welding steel, especially when taking separate heats, as the steel will unite very readily and not slip away as is sometimes the case when using plain borax. When using the Climax Welding Compound be sure and follow the directions given by the manufacturer and also the instructions that I have given in this book in reference to the welding heat of steel.

Hardening Compounds.

There are many kinds of compounds used for hardening steel and most of them are of no value. Some blacksmith will have a certain compound which he says will toughen the steel, another will say he has something that will improve the steel, while others think, no matter how the steel is worked or heated if it is only dipped in some wonderful liquid kept in a fancy pail or bucket the steel will be all right, and some have said to me in my travels, "If I knew what you had in the bucket I could make the tools stand as well as you." To explain, I happened to be in a country village and asked the village blacksmith to let me have a fire which he was not using, as I had some mill picks, axes and chisels to dress. Well, I went to work and put some salt into a bucket of water, but as I wanted to have a joke on the blacksmith, I had the salt in different sizes of paper bags, so of course as I was emptying the different papers the blacksmith and his apprentice were watching me very closely, and as they had heard I was coming to the shop they wanted to learn all they could. However, I went to work and dressed the tools, occasionally showing the blacksmith what they would do, then I went away and left the brine in the pail as I was not expected back again. The next day the story was circulated that I went away and left the mixture in a pail, and the village blacksmith could temper tools now as well as I, consequently some of his customers heard of it and they were taking him some tools and he soon had a large number of tools to dress. But a few days after I happened to be in the same village and so I called into the blacksmith shop to have a conversation. The blacksmith not expecting me, I caught him at work on some tools that had just come in to be dressed, and after taking a look around the shop I saw some other tools that the blacksmith had dressed, but they were returned to him to do over again, as they were broken. Not only were the tools broken and giving poor satisfaction, but the blacksmith was getting himself in hot water by spoiling his customers' tools and trying to do something which he did not understand. This **may** look like a fish story to the reader, but nevertheless it is quite true and I could relate other such happenings.

Now, reader, there is positively no witchcraft or common sense connected with hardening compounds, as they neither toughen or improve the steel; not only are they of no value, but the blacksmith would need almost a drug store to mix some of them, while the blacksmith works too hard for his money to spend it on such rubbish. The best hardening compound is simply clean cold water and salt to form a brine; the water should contain as much salt as the water will soak up or dissolve. This is the best compound yet discovered to harden steel at a low heat, and, reader, bear in mind that the lower the heat which steel can be hardened at the tougher it will be, and this is one of the greatest secrets connected with toughening steel. Keep the hardening bath as clean and as cold as possible. Water and brine are the only two hardening compounds used by the author.

How to Determine the Temper of Tools.

As much information could be given regarding the shape of the tool as the temper, and if tools are to be made to cut or work on material that is not mentioned in this book, it would be best to find out what hardness the material is also if the tool is to do its work by steady pressure as a lathe tool or by a blow from a hammer as a cold chisel. If the tool is to do its work by striking it with a hammer, it must be ascertained how heavy the blow is to be. For example, if the tool is struck very lightly, although the tool is to chip very hard material, it can be drawn out very thin, but if the blow is very heavy the tool must be made heavy and thick accordingly, to stand the force of the blow.

When making tools to work on strange material, it will be best to make them on the thick and temper them on the soft side, after which the hardness of the temper can be increased and the thickness of the tool reduced to whatever gives the workman (who is to use it) the best satisfaction. If it should be otherwise by having the tools drawn out too thin and the temper left too hard, causing the tools to break very easy, there will be a good many chances to one if the blacksmith or tool dresser does not lose his job, so make sure and be on the safe side in the beginning. To more fully explain: I have known blacksmiths to take jobs sharpening granite cutters' tools, and although the tools were hardened and tempered first class they were drawn out too thin, consequently the tools were easily broken, as the hammers used by ordinary granite cutters are very heavy, and likewise the blows that are struck upon the tools "are very heavy," and as granite tools require a hard temper in order to cut or chip the stone, the only method to fall back on for safety is to increase the thickness of the tool by not drawing it out so thin.

There are times however when it is quite a difficult problem to determine the correct temper. Take for example, the toolsmith who is making or dressing the chisels in a large locomotive shop, when a large number of chisels is brought to him to be dressed at one time, he does not know if a certain chisel is going to be used in the erecting department or at the motion bench. A chisel used in the erecting department has to stand a great amount of rough usage by being used for a wedge, but does not chip any hard material more than splitting iron nuts, this chisel when dressing (and its use is understood) should not be drawn out so thin as a chisel that is to do fine chipping, and it may be classed as the ordinary chisel (see chisel No. 2, Fig. 16, shown in another chapter of this book). The temper should be drawn to a very light blue almost a grey. But a chisel used by a machinist at the motion bench can be drawn out very thin (see chisel No. 1 of the same figure as just previously mentioned). The reason why this chisel can be drawn out so thin, is because the machinist as a rule is an expert in using a chisel, as he strikes it squarely on the head and holds it firm to the material he is chipping, consequently the temper "can be left harder" without fear of the chisel breaking.

Overheating Tools.

If at any time tools of a flat surface such as cold chisels, axes, etc., become overheated when hardening, never attempt to quench the tool in the hardening bath while the steel is at such a high heat but rehammer it at a low heat equally on both sides, then the tool is all right again to be heated to harden. Should the tool be quenched or hardened at such a high heat, it is very apt to crack while hardening or it will break very easy when in use. Tools of irregular shape such as milling cutters, taps, dies, etc., cannot be worked over again with the hammer, consequently great care must be exercised when heating to harden or the tool will be ruined, but if the tool is not heated hot enough to harden the first time no harm is done and it can be heated again to a little higher heat.

Cutting Steel When Cold.

Cutting the steel cold is a very satisfactory method, when bar steel is to be cut or broken into certain lengths, as when making cold chisels or other similar tools but the advantage of this method will cease when cutting steel over a certain size. For example, octagon, round or square cast steel ranging from the smallest size up to 1¼ inches in diameter, can be broken very quickly and satisfactorily when perfectly cold, by nicking the bar equally from all sides, afterwards placing the nicked part of the bar directly over the square hole of the anvil, then striking it with a sledge when it will break. But care must be exercised when breaking steel after this method, as the pieces are very apt to fly and strike the blacksmith or his helper, but to overcome this danger place the handle of the chisel on the piece which is to be broken off before striking it with the sledge, which will prevent the piece from flying. When nicking the steel, hold the chisel so as to cut in a straight line and so enable the steel to break off square on the ends.

To enable the steel to break with greater ease pour a little cold water directly on the nicked part of the steel. By pouring cold water on the steel all the heat is taken out, as steel will break more readily when perfectly cold than when it is warm.

Breaking a bar of steel cold, is a very good way of finding out the hardness or the quality of the steel, for example take a bar of ¾ or ⅞ steel (after being nicked) if the steel breaks with one or two blows from the sledge it denotes hard steel, but soft steel will require five or six blows before it breaks; also hard steel (by looking at the break) will show a fine and close fracture, but the fracture of soft steel will be more coarse and rough. If the steel is of good quality, the break or fracture will show a very uniform and silvery white appearance clear through the bar, but if the steel is of a poor quality it will show a dull brown appearance.

To test steel bars that are too large to break cold, for example a bar two inches in diameter, heat the bar to a deep cherry red, then cut in from all sides say half an inch deep with a hot chisel, then lay the bar down to cool and when it is perfectly cold it may be broken by striking it with a sledge or dropping the bar over the anvil and the quality or hardness can be judged as formerly explained. But bear in mind that the steel must not be heated above a deep cherry red (in order to cut in the nick) or the fracture when broken cannot be judged correct, as a high heat in the steel would materially change the appearance and form of the fracture. I also wish to add, when cutting a great amount of cold steel at one time, dampen the cutting edge of the chisel with oil; again, if used with care the chisel may be tempered to a purple without danger of breaking if made from steel of the proper hardness.

Oil Tempering.

Oil tempering, although often talked about, is a process little understood by the average blacksmith or steelworker and a great many mechanics have the idea that oil tempering is simply cooling off the steel in oil after the temper has been drawn. But to those whose knowledge is very limited as regards oil tempering, I will give the process, thus: In some large tool factories where tempering by colors is done away with, the temper is drawn on the tool after hardening by placing it in a vat of heated oil, the temper can be drawn to any degree according to the degree of heat the oil is heated to, which is registered by a thermometer attached to the vat.

Oil tempering does not refine the steel in any way as some mechanics think it does, but it has an advantage in this way, the temper can be drawn very evenly to any degree, also when drawing the temper in oil the steel does not have to be polished.

Drawing the Temper over the Open or Blacksmith's Fire.

The method of drawing the temper on tools over the fire is a very useful one, although a great many blacksmiths or tooldressers are not acquainted with it, as they think the only way is to let the temper run down on its own accord. I have already explained the method many times in this book, but there are a few ideas I wish to add. Of course the work that can be accomplished by this method is limited to a certain class of tools; take for example tools that are partly hardened, such as cold chisels, axes, or any similar tools.

Very often a tool is dipped a little too deep in the hardening bath, consequently there isn't enough heat left in the tool to allow the temper to run down on its own accord to the desired color, and so the necessary temper must be drawn over the fire if the best results are to be expected.

When drawing the temper over the fire be careful not to have too much blaze (better still to have no blaze), and do not have a smoky fire if it can be avoided, but in case the fire is smoky have a piece of cloth made stiff by winding around it a piece of fine, pliable wire, so that when drawing the temper, occasionally brush the tool with the cloth (where the temper is to be drawn), which will take off the smoke and keep the tool bright, and also allow the temper to be seen and drawn with greater ease and exactness. Again, when drawing the temper over the fire, do not hold the tool too close, but hold it about two inches above the surface of the fire. Also bear in mind, do not give the fire too strong a blast (just enough to keep the fire bright is plenty), as it will draw the temper too quickly on the extreme cutting edge first. Do not hold the tool perfectly still when drawing the temper as there may be a hotter spot in one place in the fire than another which would draw the temper in streaks, but move the tool sideways or lengthways back and forth whichever is best to suit the shape of the tool, until the temper is drawn very evenly.

If one side of a tool is seen to be drawing the necessary temper first, lower the other side nearer to the fire; this information will apply more especially to tools having a wide and unequal shape. Also when drawing the temper on tools, such as round punches or any similar tool that are partly hardened, keep the tool slowly and continually revolving around, in order to draw the temper very evenly on all sides. Otherwise if a round tool is held perfectly still over the fire, one side will draw to a blue while the opposite side will only heat to a straw color, unless the tool is very small.

More Points on Hammering Steel.

When hammering steel in the finishing stage to refine it, bear in mind to forge the tool as near the shape or size as possible while the steel is at a bright yellow heat, but leaving the tool a trifle large or thick, as the tool will naturally require a little stock in order that the tool will be the right width or size after being hammered. This information will apply more particularly to tools of a flat shape.

Too much hammering is not good for steel, for example, supposing the toolsmith has a flat cold chisel to draw out. He draws out the chisel and hammers it in the finishing stage to refine the steel. Now after the chisel has been drawn out to a certain thickness and hammered sufficiently, the toolsmith decides he will have to draw the chisel much thinner as he has been informed the chisel is to do some very fine chipping. But I want to say to the toolsmith right here, do not draw the chisel thinner by continuing to hammer it at a low heat, because it will have a great tendency to cause strains in the steel, which would result by cracking when hardening. But to draw the chisel thinner, heat it again to a bright yellow heat, then draw out to almost as thin as is required, then finish by hammering it as before.

Do not attempt to bend cast steel at a dull red or black heat after it has been hammered, as that would destroy all the tenacity put in the steel by hammering. If the steel is to be bent after it has been hammered, heat the steel to a cherry red heat, although a certain amount of the tenacity will be destroyed, yet not enough to injure the quality of the steel. But if the steel should be hammered, then bent at a black heat or otherwise should it be bent at a white heat, then all the toughness has been taken out of the steel. This information will apply to fine flat springs that sometimes have to be bent to the correct shape after the hammering has been done, and will also apply when bending lips on flat drills.

The correct heat for hammering steel so that it will be refined and made tough, is a dull red heat, but do not hammer steel after it becomes black, because if the steel is hammered after it becomes black it will be brittle and flakey, which will cause the cutting edge of the tool to break more easily when put in use.

Very often cold chisels or similar tools are brought to the toolsmith to be hardened and tempered which were forged by another mechanic whose knowledge concerning steel was very limited. Now "if the best results are expected" do not simply harden and temper the chisel, as the chisel must likewise be hammered. But the toolsmith will say the chisel is already drawn out thin enough, but we will suppose it is, heat the chisel to a bright cherry red heat and upset it a little, which will give a little stock to allow it to be hammered to the right size again, also, by upsetting the chisel will help to take out any light strains which may have been put in the steel by the man who forged the chisel.

Again, when hammering steel do not use too light a hammer as it is only time lost. A hammer weighing two pounds is plenty light enough, and when hammering cold chisels (unless very fine) always strike level and as hard as you can, in order to pack and refine the steel sufficiently. Bear in mind when hammering tools that have a flat surface for the last time, never strike one blow edgewise but strike every blow on the flat surface and both sides the same.

How to Improve.

Improvement is the gateway of true success in every art, trade or profession, and which applies especially to the toolsmith. To improve, the toolsmith must be devoted to his work and give it his whole mind and attention, as no toolsmith will ever be a successful steelworker, if he allows himself to become discontented by thinking that some other business or trade is better than his own, or if he only works at his own trade to make a living, and consequently all the time looking for quitting time and payday.

To the toolsmith who is determined to improve and be in the front rank, I wish to give this advice, first of all read and study this book from beginning to end, do not simply read it once but read it many times until you have the contents almost by heart, and put the instructions into practice. If you are making or dressing any tool that is mentioned in this book and it should happen to break or in any other way not give satisfaction, somewhere in these pages you will find the cause of your trouble. But if you are making a tool not mentioned in this book and it should break, find out the reason it broke, also find out how the tool does its work and if there is a weak point in the tool. Very often a tool will break, although it is made from the best quality of steel and it is properly forged, hardened and tempered, and so always remember to make a tool that is apt to break as strong as possible in every way, and do not temper the tool any harder than "just enough to do the work."

Again, when trying to improve do not accept the advice of every Tom, Dick or Harry as being the best way, without first giving it a trial, no matter if the advice "does come" from the foreman, superintendent or the master mechanic. The manager of the company whom you are employed with may be competent to run the business successfully, but remember, in 19 cases out of 20 he knows nothing about steel from a practical standpoint, with the exception of what he has been told. The toolsmith who takes everybody's advice without giving it a trial by trying to please everybody, will never improve or meet with success. To improve and become an expert toolsmith, learn all you can about working steel, as it is better in these days of great competition to be master of one trade than a Jack of all trades. Never say to yourself I can't do this job or I can't do that, but go ahead and try; do your best and if you fail, always try again until you accomplish the work in a first class and satisfactory manner. Bear in mind that success is reached by overcoming difficulties. The author has had many a hard trial and difficulty to overcome connected with steel, but by hard work, deep study and perseverance has been crowned with success.

The Blacksmith's Helper.

A willing and intelligent helper is a great help to any blacksmith, and very often the work can be accomplished with greater ease and quicker than if a blacksmith has a don't care and a dull minded sort of a fellow for his helper, and for the tool fire the helper should be fully up to the average for intelligence.

I know helpers who have better ideas concerning how the work should be done than a great many blacksmiths themselves, and very often a blacksmith has been greatly indebted to his helper for certain ideas. No blacksmith should have a helper whom he has got to be teaching or telling all the time how to strike, neither should a helper be helping an overbearing blacksmith, as I have known some blacksmiths to be changing helpers every week or two, because the helper would rather lose his job than help a blacksmith who was continually using him more like a machine than a brother shop mate. I have had helpers helping me on the tool fire from one year's end to the other, and

I always treated my helpers as I would like to be treated myself and often forming a close and lasting friendship. And to all my brother mechanics I wish to say, treat your helper as you would like to be treated if you were in his place. Do not use him like a slave by making him do heavy striking when it is possible to do the work under the steam hammer.

The Danger of Heating More of a Tool When Dressing Than What is to be Forged or Hammered.

To explain this subject fully, we will suppose a cold chisel is heated to a high yellow or white heat two inches back from the cutting edge, but it is only forged or hammered one inch back of the cutting edge. Now if this chisel should be hardened two inches from the cutting edge, it would break very easy just back of the hammered part, when put in use, for this reason. If steel is once heated to a very high heat and not forged or hammered but hardened, although it should be hardened at the proper heat, it does not become crystalized the same as when forged or hammered. Also, when the steel breaks at the unforged part, the break will present a very coarse fracture resembling a piece of overheated steel, so bear in mind to forge or hammer all the steel that is heated to a high heat, especially if it is to be hardened. If the steel is not to be hardened, it is not necessary to be so particular in working all the heated steel, although steel is always stronger when finished at a low heat whether it is to be hardened or left unhardened. This information will apply directly to small granite hand drills, where only half an inch back of the cutting edge can be worked with the hammer, and so when dressing small granite

218   THE TWENTIETH CENTURY

hand drills (or any similar drill) be careful not to heat to a high yellow any farther back from the cutting edge than 3/8 of an inch. But when dressing a large hand drill such as a miner's hand drill, it will have to be heated according to the size.

Hardening Very Small or Thin Tools.

When hardening very fine tools, have a small can of cold water or brine placed as close as possible to the fire, then the tool can be quenched immediately after it leaves the fire. Otherwise very thin tools will not hold the necessary heat (which is required to harden them successfully) long enough to reach the ordinary hardening bath.

More Information About Cold Chisels.

Although I have given more information concerning a cold chisel than any other tool mentioned in this book, I have done so because there is no one tool which requires so much science or skill as a cold chisel. I know some toolsmiths who have worked steel for forty years and yet never learned to make a cold chisel that would chip any material harder than ordinary cast steel and even then it was only guess work. Now, reader, I want to impress deeply on your mind, that cold chisels can be made to chip from the softest known material up to the hardest of chilled metal, but the chisels to do this work successfully must vary in softness or hardness of temper according to the hardness of the material to be chipped, and also vary in shape according to the weight of the blow struck by the hammer. There are a great many mechanics who think there is some way of making and tempering a cold chisel so that it will chip everything without breaking or being too soft. This is a great mistake, as it is impossible for one cold chisel to chip every kind of material and at the same time give satisfactory results. Instead we must have a number of cold chisels and each one made for a certain use.

The different shapes of chisels as illustrated in another chapter of this book, will be found to give the very best results if made according to the instructions and each chisel used for its own particular purpose. Blacksmiths and ordinary tool dressers as a rule never take the shape of a chisel into consideration; consequently, they will very often make a chisel very thin when it is required to be made thick or vice versa. When making chisels or any other tool, bear in mind that no matter how good the quality of the steel may be or how well it may be worked, hardened or tempered, the strength of the tool is always limited, consequently a thin chisel will always break much easier than a thick one especially when given hard and rough usage. Of course the general or country blacksmith who is making chisels for farmers and others, cannot tell how the chisel is going to be used or what material it is to chip, therefore the chisel must not be made too thin or too thick; it must be medium, which I have classed as "the ordinary or farmer's chisel," and which should be tempered to a light blue.

Never temper a cold chisel above a light blue unless you know for certain it is to chip very hard cast steel or cast iron. A chisel to chip very hard cast steel of about one per cent carbon, should be tempered to a dark blue. The cold chisel No. 4 as illustrated in Fig. 16 which will be found in another chapter of this book, is used for exceedingly hard and rough chipping, the shape will also apply to long chisel bars such as are used in the erecting departments of locomotive shops, and a chisel bar for this particular purpose should be tempered to a very light blue or almost a grey, as a chisel bar does not have to cut or chip any hard material, but it is given very rough usage, consequently, it must be made with a very short taper and tempered very low in order to keep it from breaking.

Always remember that if a chisel (or any other tool) is properly hardened, the chisel will stand first class even if the temper is not drawn to the exact color. But if the chisel is improperly hardened by being overheated, it will never stand or do good work no matter what temper is drawn afterwards, so make sure the chisel is properly hardened.

Although steel of 75 point carbon is best for making all kinds of cold chisels, on account of some bars of octagon steel being much higher in carbon than others it is almost impossible to always make chisels from steel of the proper hardness, and so I wish to say to the blacksmith or tooldresser, any time you have to dress or make a cold chisel from very high carbon steel (say one per cent) harden it at as low a heat as it will harden at successfully, and always let the temper run down lower. For ordinary use, let the temper of a chisel made from high carbon steel draw to almost a grey and it will give good results, but bear in mind never make a chisel from high carbon steel when it is possible to make one from steel of the proper carbon, for this reason, a chisel made from high carbon steel will keep breaking or splitting off at the end which is struck by the hammer. When dressing cold chisels, always cut off the old cutting edge after the chisel is drawn out to the right thickness before hammering for the last time, then file or grind on a new cutting edge. In the ordinary blacksmith shop the cutting edge is filed on before tempering the chisel, but in large machine shops the cutting edge is ground on after the chisel is tempered.

The different degrees of temperature Fahrenheit required to equal the various colors when drawing the temper in hot air or oil:

Color.	Deg. of Tem. F.
Light straw	440
Dark straw	470
Copper	500
Red	520
Purple	540
Dark blue	560
Light blue	590
Grey	620

Table of ordinary tools made from cast steel, arranged alphabetically, giving the color of temper and about the percentage of carbon the steel should contain to give the best results. To understand the following table of carbon, I will explain, 0.75 is equal to 75 points, 1.00 is equal to 1 per cent, 1.25 is equal to 125 points or 1¼ per cent.

Description of tool.	Color of temper	Carbon
Axe, broad	Light blue	0.75
Axe, lumberman's chopping	Light blue	0.75
Axe, limestone tooth	Light blue	0.75
Beading tool, boilermaker's	Light blue	0.75
Calking tool, boilermaker's	Light blue	0.75
Canthooks	Light blue	0.75
Centers, lathe	Purple	0.90

Description of tool.	Color of temper	Carbon
Chisel, machinists' cold	Light blue	0.75
Chisel, ordinary or farmers' cold	Light blue	0.75
Chisel boilermaker's cold	Light blue	0.75
Chisel, blacksmiths' hot	Light blue	0.75
Chisel, blacksmiths' cold	Light blue	0.75
Chisel, railroad track	Light blue	0.75
Chisel, limestone	Light blue	0.75
Chisel, sandstone	Light blue	0.75
Chisel, ordinary granite	Light straw	0.75
Chisel, marble	Very light straw	0.75
Chisel, carpenters'	Dark blue	0.75
Chisel, brick	Light blue	0.75
Clamp, bolt	Light blue	0.75
Cleaver butchers'	Light blue	0.75
Clippers, blacksmiths' bolt	Light blue	0.75
Clinch cutter, horseshoers'	Light blue	0.75
Cutter, ordinary milling	Dark straw	0.90
Cutter, pipe	Purple	0.75
Cutter, horseshoers' hoof	Light blue	0.75
Die, ordinary threading	Dark straw	0.90
Digging bars	Light blue	0.75
Drill, twist	Purple	0.90
Drill, ordinary flat	Purple	0.75
Drill, soft rock well	Dark blue	0.75
Drill, small granite hand	Light straw	0.75
Drill, limestone hand	Light blue	0.75
Drill, limestone ball	Light blue	0.75
Drill, sandstone	Light blue	0.75
Drill, small marble	Very light straw	0.75
Hammer, granite bush	Light straw	0.75
Hammer, limestone bush	Light blue	0.75
Hammer, ordinary granite	Light straw	0.75
Hammer, machinists'	Light blue	0.75
Hammer, blacksmiths'	Light blue	0.75
Hammer, car wheel inspectors'	Light blue	0.75
Hardy	Light blue	0.75
Hatchets, woodworkers'	Light blue	0.75

Description of tool.	Color of temper	Carbon
Jaws, blacksmiths' vise	Dark blue	0.75
Jaws, pipe vise	Dark blue	0.75
Knife, pruning	Light blue	0.75
Knife, butchers'	Light blue	0.75
Knife, pocket	Light blue	0.75
Knife, draw	Light blue	0.75
Knife, horseshoers'	Dark blue	0.75
Knife, carpenters' plane	Dark blue	0.75
Knife, harnessmakers'	Dark blue	0.75
Maul, railroad spike	Light blue	0.75
Pick, dirt	Light blue	0.75
Pin, flue expander	Dark blue	0.75
Pincers, horseshoers'	Light blue	0.75
Pitching tool, limestone	Light blue	0.75
Pitching tool, sandstone	Light blue	0.75
Planer tool, soft stone roughing	Light straw	0.90
Planer tool, ordinary machinists'	Copper	0.90
Pliers, wire	Light blue	0.75
Point, granite	Light straw	0.75
Point, limestone or sandstone	Light blue	0.75
Punch, boilermakers' hand	Light blue	0.75
Punch, nail	Light blue	0.75
Punch, boilermakers' machine	Light blue	0.75
Punch, saw gumming	Dark blue	0.75
Punch, ordinary center	Dark blue	0.75
Razor	Purple	0.75
Reamer, ordinary	Purple	0.90
Rolls, flue expander	Dark blue	0.75
Scraper, wood	Purple	0.75
Screwdriver	Grey	0.75
Set, nail	Light blue	0.75
Set, bricklayers'	Light blue	0.75
Shear blades	Dark straw	0.75
Snap, boilermakers' rivet	Light blue	0.75
Spring, gum (see springs as illustrated)	Very light blue	0.60

Description of tool.	Color of temper	Carbon
Spring, trap (see springs as illustrated)	Very light blue	0.60
Tap, ordinary threading	Dark straw	0.90
Tool, woodturners' lathe	Dark blue	0.75
Tool, ordinary machinists' lathe	Copper	0.90
Tool, stone lathe	Copper	0.90
Wrench, alligator	Dark blue	0.75

Table of tools continued, which are partly or wholly hardened but have no temper drawn.

Description of tool.	Carbon
Chisel, cold, for excessive hard metal	0.75
Chisel, stonecutters', for very hard granite	0.75
Cutter, milling, heavy formed	0.90
Cutter, milling, for hard metal	0.90
Dies, forging machine	0.75
Drill, large hand, for granite or hard rock	0.75
Drill, miners' cross, for granite or hard rock	0.75
Drill, well, for hard rock	0.75
Drill, machinists' flat, for very hard material	0.90
File	1.00 to 1.25
Hammer, stonecutters' mash	0.75
Hammer, ordinary stone	0.75
Pick, mill	0.90
Reamer, heavy tapered, for hard material	0.90
Reamer, for granite or hard rock	0.75
Rasp	1.00 to 1.25
Scraper for cast iron or steel	0.75
Scratchawl for cast iron	0.75
Share, plow	0.75
Stake anvil, for dressing stonecutters' tools	0.75
Steel, butchers'	1.00
Tool, machinists' lathe, for hard cast iron	0.90
Tool, machinists' planer, for hard cast iron	0.90
Tool, granite pitching	0.75
Tools, stone carvers' fine	0.75

Working Steel at Night.

Considering that the author has done a great deal of toolmaking at night, the tools have never been of such a good quality or given such good satisfaction as when made in daylight. Although there is a certain class of tools which can be made with greater success than others, for example take miners' tools. These tools are as a rule hardened but not tempered. Therefore the heat in the steel can be seen more clearly at night than the different colors of the temper. Drawing the temper on tools after night is very hard on the eyesight and even then the correct color is very often guessed at. When tempering by colors (if possible at all), arrange your work so that the tempering may be done while there is good daylight, and this rule will apply more especially to the toolsmith when the days are short if the best results are required. Too much light at the tool fire is not good. When there is too much light have a blind put up at the window which will act as a shade and which will apply more especially when the sun is shining directly on the fire.

A Few Words in Reference to Burnt Steel.

There are a great many steelworkers, who are always looking for some method or compound to restore burnt steel and so for the reader's benefit, I will give the following information: The meaning of "burnt steel" is steel that has been heated to a higher heat than what it would stand, thus, when the steel is burnt it falls or flies to pieces when being struck by the hammer.

The best method the author has yet discovered is,

being careful in the first place not to burn it, as an ounce of prevention is worth a pound of cure. The next best method is, cut off all the steel that is burnt, as it is only time lost trying to restore it to its natural state, and the time lost is of more value than the steel. Supposing it should be restored successfully, but bear in mind, that steel when once burnt is never as good as the steel in its ordinary natural state.

Special Methods of Hardening Tools and Steel Goods.

The following remarks may serve as an amplification of the general principles already enunciated, and as an explanation of the special methods. The processes to be described have justified themselves, although it cannot be said definitely that other methods might not serve as well. In considering the question I shall pass from partial to total, from the simpler to the more difficult hardening, and shall allude to the hardening of certain steel goods for the sake of completeness.

File-Cutters' Chisels.

These are hardened in water to a depth of from 15 to 20 mm., according to their size, after being raised to a dull red heat, which must be uniform over the whole width of the edge, which is the only part hardened. The chisels are tempered at a purple red on red-hot iron. They can be ground down about 8 mm., after which fresh hardening is wanted.

Before this rehardening the edge must be forged thinner, as it has already become too thick by wear and grinding. At this time, too, the shank is forged out smooth again. If the shank has been hardened it might fly, to the danger of the eye or hand of the file-cutter. Hence it is a mistake to believe that a hardened shank of a file-cutter's chisel lasts longer than an unhardened shank.

Paper and Tobacco Knives and Shears.

Long articles of these kinds are best heated in a cupola and kept there till the edge has reached the hardening temperature. These cutting tools have generally holes and slots in them whereby they are fixed into the machine. So that the thin edges of these apertures may not get hotter than the solid metal near them, they should be stopped with loam before heating. The hardening is done in water or, in difficult cases, in tallow, into which the knife is put horizontally with its back outwards.

For tempering we use a rectangular iron frame, somewhat longer than the knife. Several holes are cut below in the sides of the frame, and fire-bars rest on bearers a little way above the holes. This permits the access of air to the coal-fire which is made in the box. When a sufficient glow is got, two thin iron bars are laid across the box and the knives are laid on them with their backs downwards. There they remain till their edges appear violet. Another way is to lay the knives flat, so that only half the width of the lowest knife is over the fire. The second half covers the first, the third the second, and so on, so that only the backs of the knives are exposed to the direct heat of the fire, the edges being heated by conduction only. The reason is that the edge must be hard and the rest of the knife soft and tough.

Short knives are dipped vertically in the hardening liquid.

Short thick shear-blades are heated with the cutting side downwards in an open forge fire, only at the edge, but not bounding the heated part too closely. The edge is then dipped horizontally into the hardening liquid, the process being similar to that used for chisels and plane irons.

Stamps.

Stamps can be classified from their chief characters as follows:—Punches used for making holes in metal. The shape and size of the flat ground surface corresponds to, and determines the shape and size of the hole made, and of the piece of metal punched out. Stamps are matrices worked by the hand, like a chisel. They consist of short steel rods with working faces of the proper form, and are used principally for making hollow bodies and for inscribing letters on metal, and ornamenting it by inlaying. Strictly speaking, the already described stamping hammers come into this class. In many cases, the gradual action of a large number of weak blows with the hammer may be advantageously replaced, with the aid of a machine, by a single powerful impulse or pressure, in which case the form to be given is determined by a matrix or die. Stamps include both matrices and relief dies. For stamping metal buttons, jewelry, ornaments, etc., matrices are generally used (rarely relief dies), and are called stamps most generally. The stamp has an under and an upper part. In coining, two stamps are used, the lower one being fixed, while the upper one comes down vertically above it. A ring confines the edges of the coin during the stamping so as to prevent the deformation of them. The second stamp, made from an original stamp, is called he matrix or model-stamp, and is used for making dies by pressing.

1. Minting Dies.—These are heated red hot in charcoal in an iron box smeared with loam, but not closely shut. The working face of the die is then hardened by allowing a stream of water to fall upon its centre, as it is difficult to get that part as hot as the edge, and yet the whole surface must have a uniform hardness. The working surface is usually not tempered, and the rest of the die only a little, so that the die may not be deformed in use by the powerful pressure.

2. **Press Tools.**—Such dies with very fine engraving are hardened like coining dies. Those for button- and ornament-making and also for rivet manufacture can, however, be heated in an open charcoal fire. They are heated up, slowly at first, till the whole mass, and particularly the working surface, appears of a uniform brown red. Then the working surface is quickly brought to a higher temperature, as uniform all over as possible, and shading off into the brown red of the rest of the tool. For this purpose the fire must be kept short. To compensate for the unavoidably unequal heating of the working surface, the hardening should be done with a stream of water falling on the middle of it. If the edge gets harder than the rest it must be tempered rather more to prevent cracking in use. The tempering is done in a sand bath, or an iron plate floating on fused lead, the working surface being upwards so as to temper the body of the tool the most.

To prevent the tempering color from getting too much into the middle, that part is cooled with drops of water. For press tools used with heavy blows, tallow hardening is advisable.

3. **Drawing Rods for Cartridge-Case Making.**— These are hardened over their whole length, and there is the danger to be encountered that from want of care they may be warped. They are best heated inside a

uniformly heated tube lying in a long bedded forge fire with several tryers, or in a muffle. The hardening consists in a vertical dipping into water until the surface of the steel is dark, and then finishing the cooling in oil or tallow. Tempering is usually dispensed with.

4. **Matrices for Rivet-making.**—These must be hard on the upper surface, on the inner edge of this surface, and also in its anterior walls, but very tough and resistant to blows in the rest of its mass. The heating has therefore to be done accordingly, which is difficult. After a preliminary gentle heating, as uniform as possible, of the whole tool, it is brought into the hot part of the fire with its working surface downwards, so as to give that the full hardening heat. The quenching is best done with a stream of water, directed not merely through the opening of the matrix but over all its upper surface. If the outer edge scales fairly completely it must be tempered yellow, which can be done with a red-hot iron ring. Matrices for cartridge-case manufacture are quenched the same way.

5. **Short Stamps with large Working Face.**—These are heated entirely and uniformly to the hardening temperature, and quickly cooled in a large vessel of water, with a constant flow through it, by dipping vertically and stirring them about. If possible the flow of water into the reservoir should be under pressure from below, in which case the stamps can be held quietly in the current. This is enough for soft steel (about 7 per cent. of carbon), but for harder steels the quenching in water is followed by a cooling in oil. In neither case is the steel tempered. If the stamps are made of medium or of hard steel, they must only be hardened in oil or tallow, and must be slightly tempered. Soft steel should be preferred, as it is more

easily engraved. In heating such steel the working surface should be sprinkled with ferrocyanide of potash, both to prevent oxidation and to get greater hardness. The residual ferrocyanide is quickly removed with a fine wire brush before the hardening temperature is reached, and renewed thinly and uniformly by means of a wire sieve.

6. **Long Stamps with large Working Surface.**—These are warmed all over, but the further heating is only at the working end, the cherry red passing into a brown red a little way up the tool. The quenching is done, like that of hammer heads, in water with a rapidly rising inflow. The overflow pipe must be so arranged that the surface of the water does not come more than about 2 cm. above the working face of the tool.

The chief difference between the hardening of short and long stamps is, that the former are hardened all over, the latter only at the face. If short stamps were only hardened partially they would warp, and this need not be feared with long stamps, as the larger mass of steel proceeding from the working face opposes a sufficient resistance to warping.

7. **Cutting Matrices for punching out Plates for Tinman's Work, etc.**—These are heated like the stamps for rivet-making, and quenched by vertical dipping in water, or for hard steel in oil or tallow. Tempering colour: yellowish brown to brown red.

In general, it may be remarked that with stamps which have to work with great accuracy, and which must therefore be of an exact size after hardening, the change of volume produced by the hardening must be allowed for. Hence a solid tool must be made somewhat too small, a hollow one somewhat too large, and just enough to compensate for the expansion in the one case and the contraction in the other. As the change in volume is different for different kinds of steel, it must be determined beforehand for each kind.

Saws.

Haswell recommends the following hardening method for saws in Karmarsch and Heeren's technical dictionary: "Circular saws are brought to a cherry red and quenched in water with a thin layer of oil on its surface. The heating must be done slowly. The saws are immersed vertically. The oil catches fire as it touches the hot steel, and covers it with a crust of carbon, which protects it from too quick cooling and makes cracking less likely. Single saws can be given the hardening heat by laying them on a cold iron plate and then heating both together, and still better by heating the saw while pressed between two wrought-iron plates. This ensures slow heating of the saw and prevents warping. For the thinnest saws only oil is used for quenching, or a mixture of oil and tallow. This gives enough hardness. Saws of medium thickness are best quenched in solid tallow. This gives a somewhat greater hardness than oil. Very thin saw blades also get hard enough if heated red hot and cooled between two iron plates smeared with tallow. Saws for metal must be tempered at a straw yellow. This is done after polishing best by laying the saw on red-hot iron."

Holzapfel describes the hardening of saw blades as follows in the Mechanic's Magazine:—"Saw blades are heated in special long stoves and then laid horizontally with the toothed edge, or the edge to be toothed, uppermost in the hardening liquid, which is a mixture of oil, tallow, wax, etc. Two troughs are generally used.

so that when one gets too hot the other can be used while it cools again. A part of the hardening liquid is wiped from the saws to a piece of leather, and they are then heated over a bright coke fire till the grease left on them catches fire. If they have to be left rather hard, only a little fat is allowed to burn on them, but more if they are to be softer. To get spring-hardness, all the fat is allowed to burn.

"With other objects, which are thick, or of unequal section, such as many springs, two or three lots of fat are burnt off them, so as to ensure the same tempering throughout."

Thin saw blades and other small objects are sometimes brought to the hardening temperature by immersion in red-hot lead, having first, as already said, smeared them with linseed oil and soot, and dried to prevent the lead from sticking to them.

Shears.

The blades are heated uniformly to a dark cherry red reaching from the point to the rivet hole. This can be done in the open fire, first with a weak blast, until the steel begins to glow. Then the fire is left to itself, and the steel is moved about in the fire till all the parts to be hardened have received a uniform dark cherry red. Both blades are hardened together in water and tempered at a purple red or violet. It is necessary to treat the two blades together throughout, so that both may be of equal hardness, so that one will not cut the other—the well-known rule of dipping the blades vertically and slowly, points uppermost, down to over the rivet hole.

Table and Pocket Knives.

There is little special about hardening table-knives. The blade is dipped slantwise at a dark cherry red, back first, into the hardening liquid, which is usually plain water, or water covered to a depth of 10 to 15 mm. with oil. The blades are then tempered at violet or blue. Pocket-knives are taken, half a dozen at a time, in the tongs, the separate blades being kept apart in the grip of the tongs by a piece of iron. They are then heated edge uppermost over a fire or hot-iron plate. For fine knives, a fused mixture of tin and lead is used for tempering the back and a spirit flame for tempering the edge.

Scythes.

The blades are heated in a little reverberatory-furnace, a small walled flue of neatly square section, which is steeply inclined towards the chimney. The grate is in front, and in front of it is a tuyer-opening. It is preferred to fire with wood rather than charcoal, as the former makes more flame and keeps out the air better. A few centimetres above the sole of the furnace iron bearers carry the scythe blades. At first one blade is put with about a third of its length in the furnace, with its back downwards and with the point foremost. When this has been somewhat heated, other blades are put in, and the first is gradually got entirely into the furnace. The number of blades being heated at once depends on the temperature of the furnace. The slow pushing in of the blades is necessary, because the greatest heat is in front, over the grate, so that the points of the scythes would get less heated than the rest if they were put at once into the fire.

The quenching is in tallow, with edge uppermost.

The scythe is removed from the hardening trough when fumes cease to come from the tallow. The tallow is scraped off the still warm steel with a sharpened piece of bark. Cherry tree bark is preferred for this. The scythe is then worked about in a heap of coal ashes to clean it. It is then heated very gently and as uniformly as possible over a charcoal fire, and then immersed with a hewing movement in a trough of running water. This brings the scale partly off the front side of the blade. The rest of it is got off with an emery wheel, and the scythe is finally blue tempered. For this purpose an iron trough of proper size is lined on three sides with glowing charcoal, leaving one of the long sides free. On this side the scythe blades are put with their backs downwards on iron bearers and brought separately in turn nearer to the heat. As the tempering colour should appear uniformly over the whole blade, any defects in it are rectified over a charcoal fire with a gentle blast, heating those parts which have not been tempered enough, while the overtempering of the rest is prevented by keeping it cool with a wet cloth applied now and then. The tempered scythes are cooled in the air. If very hard steel has been used, the scythes are tempered twice, polishing the front side of the blade before the second tempering, so that the colour can be observed. A sand bath heated by waste heat or some inferior fuel is more economical than the open charcoal trough, but does not do the tempering so well.

The subsequent operations consist principally in hammering the slightly heated blades under a very rapid but light machine-driven hammer with a very small rise. This rectifies the distortions produced by hardness, and increases the hardness and elasticity of the steel. This is followed by a final adjustment of the blade to its right form by means of hand-hammering.

Wire.

According to Tunner, piano wire is hardened in Worcester as follows:—After a lead bath in an iron pipe kept red hot, the wire is hardened in a circulating oil bath. It then passes to a second lead bath, which, however, is at a temperature only just above the fusion point, and tempers the hardened wire.

Watch Springs.

An interesting machine of Kugler of Paris for hardening watch springs is described as follows by Kohn in Karmarsch and Heeren's technical dictionary:— "After the steel wire has been rolled out to the proper thinness it is coiled up on a cylinder, from which the ribbon passes through an iron pipe surrounded outside by fireproof material and having a rectangular section of about 100 mm. wide and 12 mm. high. This pipe is in a furnace heated with charcoal. As the steel ribbon slowly passes through the pipe it becomes red hot, and is drawn through a bath of oil and hardened. The heating of this bath is prevented by a constant flow of oil, the warm oil escaping by an overflow. On leaving the oil the ribbon passes between two pairs of drying rollers, one behind the other, which are suitably loaded and from which the oil flows back into the hardening bath. The band next arrives at a cast-iron plate heated by a fire, to temper the spring. Here it is also straightened by a weight, and, finally, through a polishing apparatus consisting of six emery rollers, which polish both sides of the spring. The spring is finally wound on a reel."

A similarly acting apparatus for hardening and tempering long springs is that patented by Luttger Brothers of Solingen. Here, however, the hardening is done dry, by passing the spring as it leaves the hot pipe between two cast-iron reservoirs filled with cold water, which harden it by their pressure and coldness. The reservoirs are kept cool by a constant flow of water. The pressure of the upper reservoir on the spring can be regulated by a lever.

Hollow Steels.

The method to be followed with these has been already described, but we will here mention an apparatus patented by Lorenz of Karlsruhe. It consists essentially of the following parts:—To a vertical water pipe with a conical valve regulated by a screw, a mouthpiece is fixed below, by a coupling box. The mouthpiece must exactly fit into the object to be hardened, and it is given a spheroidal shape to enable it to fit into various sizes. Below there is a vertical overflow pipe, which can be raised or lowered by a lever, and by a spring acting on this lever is pressed up against the mouthpiece. This pipe has a suitable flange to carry the hollow steel. If the hardening is to be internal only, the red-hot steel is put between the overflow pipe and the mouthpiece, from which the water then flows through it. In this it is clear that the stream of water must fit the bore of the steel, or else the supply of water must be somewhat lessened, so that the thinner stream may spread over the inside of the steel.

If the hollow steel is to be hardened first inside and

then outside, the overflow pipe receives a sort of cast-iron pan, the edge of which reaches the upper edge of the steel to be hardened. In this pan a small tripod is put for the steel to stand on. With this arrangement the water flowing into the pan through the hollow of the steel ultimately immerses it altogether. The overflow from the pan is taken away through a side opening into the overflow pipe. To drive the hardening water close against the inner sides of the steel, the mouthpiece has a spindle in its centre, which spreads out outwards into a cone which directs the water outwards. The space left by the base of the cone for the water to pass out can be regulated.

If a hollow steel is to be hardened inside and out simultaneously, Lorenz uses an apparatus of which we can form an idea by supposing that a short coupling pipe is placed over the above-mentioned pan. This is connected above to the mouthpiece, which in this case does not rest upon the hollow steel. In this we get a closed chamber in which the hollow steel is placed free on all sides. The conical spindle sends the water both inside and outside at the same time, and it flows away by openings in the chamber.

Curing Warped Steel.

This operation is, if permissible, usually very difficult, and must be conducted with the greatest care to be successful. It is generally done after the tempering. Objects which have only been slightly tempered or not at all, are made hand warm so as to slightly lessen their brittleness but not their hardness. They are then put between warm copper blocks in a screw press, and gradually brought to the proper shape and left to cool under the pressure. More strongly tempered objects of softer steel are quickly and gently hammered on the anvil with a hammer, with a very narrow and rounded face, which stands in the direction of the hammer handle. The blows are given on the concave side of the steel, where the contracted parts lie. These are extended by the hammering till the steel has its proper shape restored. The blows must fall exactly in the direction of the bending, and not across it. The face of the hammer must be, if necessary, so small that the whole length of it lies closely on the surface when the blow is struck. If the hammer is too heavy or used with too much force, its action extends over the whole section of the object, and the warping is made worse. Besides, the steel may then easily be broken.

Blue-tempered thin elastic objects, such as blades, may, while still warm from the tempering, be bent the other way. For this purpose two pegs are fixed about 80 mm. apart in the work bench. Dietlen, in Dingler's Polytechnic Journal, recommends adjusting the warped steel during tempering. "Stretch the hardened article on a piece of iron by means of iron screws, with its concave side next the iron, and heat the whole slowly over a coal-fire. When the steel begins to be yellow, it may be slowly straightened by means of the screws. As soon as it has the desired tempering colour, it is cooled with water on the side that had been convex, and will keep its shape when the screws are removed. Very slight warping may be cured by heating the concave side, and then wetting the convex side."

It hardly needs to be mentioned that only comparatively thin objects, such as files and blades, can be cured as above if warped. Compacter objects cannot be straightened, and must be rehardened, after having been brought very slowly in a slow fire to a red heat, and then allowed to cool in the air or in wood ashes. If this does not make matters right, the piece must be reforged. If the heating is too quick, the steel may easily be cracked, and if the rehardening is done without this preliminary heating, it will probably cause cracking and fresh warping.

Conclusion.

In concluding this book, I wish to remind the readers that it has been written for their interest, and the author has endeavored to give all the necessary instructions and illustrations of all the principal tools used by almost every leading trade to insure the greatest success in the art of steelwork or toolmaking. I have left nothing unwritten which I thought would be a help or interest to the readers, and remember, readers, I have written this book to improve your mechanical ability and ideas, hoping thereby to help and encourage you to strive to reach the highest rung in the ladder of mechanical success.

It is not necessary to work at blacksmithing 10 or 15 years in order to do good work or become a first class toolsmith when you have this volume of information at hand. Readers, place a great value on your leisure hours, they will be sands of precious gold to you when spent in reading this book. Do not simply read "but think as you read," and the mechanic (whether young or old) who reads and thinks in this way will be well rewarded and soon rise above his peers. No doubt this book may be the cause of many an argument and some may condemn it as being untrue, but before condemning it take my advice and put the instructions into practice, and in years to come you will often thank the author for bringing this information before you. The mechanic who reads this book without putting the information into practical use will remain in the same old rut. My brother mechanic, you will never succeed unless you are willing to branch out and accept new ideas or methods. Do not get in the habit of thinking you know all about toolmaking or that no one can teach you anything more than what you already know, as the author takes second place to none in the art of general steelwork, yet he occasionally gets a new idea or a quicker method from some inferior mechanic. The author has written this book in order to illustrate the most up-to-date methods but if the reader (after reading) still fails to put the information into practical use or even give the different methods a trial, then he will not benefit by the reading of this book, and the author has failed in his attempt to instruct him. Bear in mind that toolwork is the very best part of blacksmithing and the blacksmith or tooldresser in any machine shop, stone yard, quarry or mine who happens to be a first class steelworker, holds the respect of all and his services are always in great demand, so reader why give up all your hopes of becoming a successful steelworker? Your chances are equal to that of the author's. He had no one to give him a word of encouragement. Neither had he a book as complete as this to help him overcome his difficulties in connection with steel, so reader let your determination be to press on and overcome every obstacle which stands in your way. Make use of all the brains which God has given you and let your ambition ever be to rise and take the lead. Your success is sure if you do your best.

Do not be given to be always telling others **what**

you can do but keep your tongue quiet and your eyes open and always be on the alert to gain knowledge in connection with your trade. If you are a first class mechanic your customers will judge your workmanship and give you a reputation, and remember, reader, a good reputation is worth striving for even if you gain it slowly. The author well knows the worth of a good reputation which he has gained slowly by the combination of hard work, deep study, close observation, a vast amount of experimenting and wide travel. Reader, the contents of this book is the author's reputation, so make sure and combine the contents of this book with your own practical experience.

In drawing this book to a close, the author trusts and hopes that every reader (who is connected with steelwork) will be greatly assisted and placed on a foundation for future success. I have not merely written this book to improve your mechanical ideas and instruct you in the art of toolmaking, but I have written it for the sake of the love which I hold for my brother mechanics. I have placed my whole heart in the work in order that others may share with me in the joys of mechanical success. Some readers may think I have been rather sarcastic at times, but if I have been it was only in reference to a certain class of mechanics, in order to point out to them their mistakes and thereby illustrate the difference between the right and wrong ways and also the difference between good and poor tools.

My closing advice to the reader is, when you are making tools that have a cutting edge, make sure that they are hardened at the right heat. Hardening steel at the right heat is the most important obstacle to be overcome in connection with the art of toolmaking, for no matter how good the quality of the steel may be or how well it is forged, the quality and success of a tool will always depend on the proper heat for hardening. Again, always remember to do your work to the very best of your ability, and follow closely the old adage, "whatever is worth doing, is worth doing well," and you will soon become master of the "king of metals."

USEFUL FORMULAS.

Tempering Brass.

No. 1. Brass is rendered hard by hammering or rolling; therefore when you make a thing of brass necessary to be tempered, prepare the material before shaping the article. Temper may be drawn from brass by heating it to cherry red and plunging it into water.

To Case Harden Set Screws for Shafting.

No. 2. Melt piece prussiate potash the size of a bean on spot you want hard while it is hot and plunge into water or linseed oil.

To Case Harden any Particular Spot, Leaving Other Spot Soft.

No. 3. Make a paste of concentrated solution of prussiate of potash and then coat the spot you wish to harden; then expose to strong heat. When red hot, plunge into cold water.

To Case Harden Cast Iron or any other Iron.

No. 4. Three parts bichromate of potash; one-half part common salt. Pulverize well and mix. Heat iron to highest heat it will stand; then sprinkle on mixture and try well on both sides. Cool in water.

To Case Harden Steel.

No. 5. Use one part oxalic acid and two parts of pulverized common potash. Pulverize them well and thoroughly mix. Heat to cherry red, then roll in mixture as you would in borax, then heat again in clear fire; cool in water.

Composition to Convert the Most Impure Scrap While in Ladle to No. 1 Castings.

No. 6.   8 pounds of Copperas.
         3 pounds of Zinc.
         ½ pound of Tin.

Throw the above amount in every hundred pounds of melted iron.

Drilling a Larger Hole through Smaller Hole with Same Drill.

No. 7. With the same drill, say you want to drill a ¾ hole in piece of iron. Now you want the hole 1 in. deep, and 1 inch deeper at bottom and larger at bottom. To make this drill the ¾ hole first 1 inch deep, use a V-shaped drill, then grind the point of same drill ⅛ to one side. Don't grind it smaller and for every ⅛ you grind the point to one side, you will drill the hole twice that size larger. It will drill shoulder where larger hole begins.

Solution to Harden Cast or Gray Iron to Any Degree.

No. 8.   1 Pint Oil of Vitriol.
         1 Bushel of Salt.
         1 Pound of Saltpeter.
         2 Pounds of Alum.
         ¼ Pound of Prussic Potash.
         ¼ Pound of Cyanide Potash.

Dissolve the whole in three gallons of rain water. Heat iron to cherry red and cool in solution.

Dressing Mill Pick.

No. 9. To dress mill picks, heat to cherry red and dip points while hot in a tallow before hammering. Then to temper them:

> 2 Ounces Muriate of Ammonia.
> 2 Ounces Chloride of Potash.
> 2 Gallons Soft Water.

Heat to cherry red and plunge in solution. If too hard add more water.

To Harden Steel Rolls.

No. 10. To prevent shrinkage in side and so prevent bursting take three or four hands full of soot and a small hand full of lime in a pail of water. Heat cherry red and cool off in solution.

In tempering cast steel or any steel always use soft water, always dip towards the North, and tempering round steel, dip perpendicular. Always leave steel in water until cold through.

To Prevent Steel from Springing.

No. 11. Have some dry common soda, heat steel to cherry red, then lay hot steel in soda. Hot steel will melt the soda to a liquid. Let it remain till cool. Will find a good temper.

Hammering Cast Steel.

No. 12. We have often seen smiths spoil a chisel or mill pick by hammering it too cold. This will not spoil a thick piece of steel but will a thin piece. Better take another heat.

Tempering Bitts, Blades or Knives without Drawing Temper.

No. 13. 1 Ounce Pulverized Corrosive Sublimate.
2 Ounces Sal Ammoniac.
Two Hands Full of Salt.

Dissolve in six quarts soft water. Heat to cherry red and plunge in solution and do not draw temper. If too strong add more water.

Solution to Temper Steel to Any Degree.

No. 14. 1 Ounce of Blue Vitriol.
1 Ounce Borax.
1 Ounce Prussic Potash.
½ Pint Salt.

Dissolve all in one quart water, then add one gallon raw linseed oil and ½ ounce pulverized charcoal. Heat cherry red. Cool in solution.

Tempering, Hardening, Toughening and Restoring Steel.

No. 15. This formula for compounding the celebrated patented Mergess solution for tempering, toughening, converting low grade cast steel to higher grade and restoring burnt steel. 4 ounces of citric acid in one gallon boiling water, dissolve two minutes, then add 4 ounces of carbonate of iron, stir for a minute. Now let it stand till agitation stops, then add 6 ounces prussiate of potash, 2 ounces of saltpeter. Then make it into 12 gallons of soft water and stir in six pounds rock salt. Solution is ready. Temper same as in water. But for edge tools bring to proper color, heat slowly, dip hot steel in solution once in a while while heating.

Tempering Steel Springs without Springing.

No. 16. Heat to cherry red, then let it cool off itself. Then coat the spring with soot that will arise from burning resin, then heat evenly until the soot disappears, then immerse in linseed oil. Will make fine temper.

Tempering in Bath, Not Fire.

No. 17. For twist drills, taps, dies, small punches or such articles of cast steel you wish to keep straight; take as follows: Equal parts of prussiate of potash and common salt, put them together in an iron pot over fire when it gets to proper temperature. It will boil and become a cherry red. Put the tool in this until it becomes a cherry red. You may leave the tool in all day if you wish, for the longer the more it improves the steel. When you take it out cool in water or linseed oil, always in a vertical position. Do not draw. But for taps or dies draw to dark straw.

To Harden Cast Iron to Cut Glass or Cutting Purposes.

No. 18. 2 pounds Common Salt,
½ pound Saltpeter,
½ pound Rock Alum,
¼ ounce Salts of Tartar,
¼ ounce Cyanide of Potash,
6 ounces Carbonate of Ammonia.

Mix and thoroughly pulverize together. Apply this to surface when the metal is cherry red and plunge in cold, soft water.

Tempering Round Piece Cast Steel without Springing.

No. 19. Stir the water fast with stick. While the water is in a whirl plunge hot steel in center of whirl perpendicular. Water turning around it will keep it straight.

Tempering Drills.

No. 20. Heat to cherry red and plunge in lump of Beeswax and Tallow mixed. Not too much tallow or will make soft.

To Temper a Thin Blade or Knife.

No. 21. Cut a piece of paper a little larger than blade, then heat blade evenly, then lay the paper flat on water, lay blade on paper and press under to cool. Never mind the theory. Try it. Always dip blades to North.

Remarks When Welding Cast Steel or Any Steel.

Always weld the same way. Begin where you left off. Take one heat and the next heat begin where you left off so the dross and scales will work out. If you weld one end then stick the other end the dirt will get in center and can't get out and you can not weld it any way.

Welding Cast Steel with Less Heat.

No. 22. Mix Sal Ammonia with ten times the amount of Borax. Fuse well when pulverized. Now mix with this an equal quantity of quick lime and use as borax.

Welding Steel Bessemer Spring Axles and Tool Steel

No. 23. 15 pounds Dry Sand,
8 ounces of Powdered Sulphate of Iron,
8 ounces of Black Manganese,
8 ounces Fine Salt. Use as Borax.

Welding Cast Steel and Restoring Burnt Steel

No. 24. ¾ pound Borax,
¼ pound of Sal Ammonia,
⅛ pound of Prussic Potash,
½ ounce of Resin,
½ gill of Alcohol.

Simmer these in spider over slow fire until well chased. Then use as Borax.

Welding Cast Iron to Steel or Iron. It Will Weld Better than is Generally Known.

No. 25. 1½ pounds of Powdered Copperas,
1 quart Fine Dry Sand,
1 Hand Full of Salt.

Now make the pieces hot and while heating dip them in mixture. Throw some on in fire. When iron and steel are hot and will stand without running, place them quickly together, rub them with piece of steel or old file, drawing soft parts over each other.

Welding Cast Steel Edge Tools or Any Fine Work.

No. 26. This is the best steel welding compound in use today and is known only by a few good smiths: Dragon blood pulverized and mixed with borax until the borax looks a little pink in color. Use as borax.

Welding Steel Boiler Tubes.

No. 27. Flare long piece out, fit short piece inside the other neatly, then lay in fire. When hot enough sprinkle on welding compound. Have helper tap lightly on end of short piece, while you take light hammer and tap it lightly in fire turning all the time. Weld it all in fire.

Repairing Plows, New Shear and Laying.

No. 28. First take old plow, set it on level board. See that it measures 16 inches from floor to hitch and has 2¼ inches land. If not, while repairing bring it to that, and then it will run right. In laying shears take hammer, lay steel 2 by 5-16 and use the welding compound mentioned above. Don't make wing of shear more than 6½ inches wide. For new shares lay steel for shares on plow, make wing 6½ inches wide, cut off on land side what you don't need. Now bend wing down shape of old. Lay share piece under and weld up.

Stream Tempering All Heavy Tools.

No. 29. We will take a hand hammer for example. Take a can or keg, make a three-eighths inch hole in it; then heat hammer a cherry red; then hold peen in slack tub and let three-eighths stream pour on center of face until cool enough; then let draw to a dark straw color. If it does not draw to right color, heat eye wedge put in hole until the right colors appear. The old way of dipping in tub cools outside too fast, cracks it and makes it shelly. The new way of cooling center the fastest contracts the steel and makes it solid, and it will never crack nor sprall off.

Redressing and Tempering Old Anvils.

No. 30. Heat old anvil to draw temper; let it cool slow, plane off face, heat face to cherry red and while hot throw on face a handful of prussiate potash. Then cool as fast as possible with a heavy stream on center of face. It will be as good as a new anvil.

Oil Tempering All Heavy Bolts, Blades and Knives.

No. 31. Heat all flat pieces, knives, blades and bitts on edge. If you lay them flat on fire you will spring them. Heat to cherry red and plunge in raw linseed oil. When cool scour off edge bright. Heat a heavy iron, lay tool on, edge up, draw to dark straw color.

Tallow Tempering for Machinists, Tools and Tools Requiring Hard, Tough Edge.

No. 32. Two-thirds tallow and one-third beeswax; add to this a little saltpeter to toughen steel. Dissolve all and mix. Heat point of tool cherry red; dip point of tool in solution as you would in water and let it draw only to a light straw color. This is a good thing. It improves the steel; all tools will have a hard, tough edge.

Case Hardening Steel Plow Mold Boards.

No. 33. Make a brine of salt and rain water to hold up an egg: add a little saltpeter. Heat steel or mold board cherry red, and while hot sprinkle on face prussiate potash and plunge toward the north in the brine. Let it lay in the brine until cool through and it will not spring nor crack.

Bending Gas Pipe without Breaking.

No. 34. Heat pipe good red heat. If heat is too long, cool off pipe to where you want the bend. Then put end of pipe in fork on anvil, and while bending let helper pour a small stream of water on inside of bend where it looks like kinking. You can bend any shape this way.

Brazing with Copper or Brass.

No. 35. Scarf the ends of pieces so they fit nice. Then clamp the pieces so they fit nice and can not slip. Then lay on fire; put on top side the copper that you think is necessary, and then put on some charred borax or Monarch Welding Compound. Then heat iron until the copper melts. Take a file and keep the copper where you want it, and then lay it down and let cool. This way you can braze iron, steel or malleable iron.

How to Weld Cast Steel with Borax.

No. 36. Put borax in a pot on a slow fire and boil it until it becomes dry like dust: Stir it all the time it is cooking. Then use the dust. You will find it welds much better, as cooking it takes the sulphur out of it, and you will get a clear fire and a nice clean heat.

How to Weld Anything Likely to Slip.

Such as steel tires, but not good for cast steel.

No. 37. To one pound of pulverized borax add two ounces of sal ammoniac. Put a little on tire cold, and when it gets hot it will get very sticky and hold the tire in place so you can handle it. When the tire gets hot put on more. Weld at a borax heat.

Welding or Soldering Band Saws.

No. 38. File scarfs so they fit together nicely; then put a piece of silver solder between laps, or a silver coin will do. Then put on some muriatic acid, or some charred borax is just as good. Then heat a pair of very heavy jawed tongs; heat to a very high heat; hold laps of saw between jaws of tongs until welded. They weld very quick, and will not break where welded. Some pour water on tongs to cool them off fast.

How to Work Self-Hardening Steel (Called Mushet Steel).

No. 39. Heat to cherry red; forge to desired shape; then heat again to cherry red; lay in air to cool—the more air the harder it will be. To make very hard, hold in cold blast.

Instructions for Tempering Pneumatic Tools.

And for some heavy shear knives where it does not require too hard a temper.

No. 40. Heat tool all over; heat very slowly, so it will heat through to cherry red, and plunge tool in linseed oil and let it lay in oil until it is cool clear through. This will give a good temper on any tool required hard all over.